A WORLD BANK COUNTRY STUDY

KOREA

Managing the Industrial Transition

Volume I
The Conduct of Industrial Policy

The World Bank
Washington, D.C., U.S.A.

The World Bank
1818 H Street, N.W.
Washington, D.C. 20433, U.S.A.

Manufactured in the United States of America
First printing March 1987

World Bank Country Studies are reports originally prepared for internal use as part of the continuing analysis by the Bank of the economic and related conditions of its developing member countries and of its dialogues with the governments. Some of the reports are published informally with the least possible delay for the use of governments and the academic, business and financial, and development communities. Thus, the typescript has not been prepared in accordance with the procedures appropriate to formal printed texts, and the World Bank accepts no responsibility for errors. The publication is supplied at a token charge to defray part of the cost of manufacture and distribution.

Any maps that accompany the text have been prepared solely for the convenience of readers. The designations and presentation of material in them do not imply the expression of any opinion whatsoever on the part of the World Bank, its affiliates, or its Board or member countries concerning the legal status of any country, territory, city, or area or of the authorities thereof or concerning the delimitation of its boundaries or its national affiliation.

The most recent World Bank publications are described in the catalog *New Publications,* a new edition of which is issued in the spring and fall of each year. The complete backlist of publications is shown in the annual *Index of Publications,* which contains an alphabetical title list and indexes of subjects, authors, and countries and regions; it is of value principally to libraries and institutional purchasers. The continuing research program is described in *The World Bank Research Program: Abstracts of Current Studies,* which is issued annually. The latest edition of each of these is available free of charge from the Publications Sales Unit, Department F, The World Bank, 1818 H Street, N.W., Washington, D.C. 20433, U.S.A., or from Publications, The World Bank, 66 avenue d'Iéna, 75116 Paris, France.

Library of Congress Cataloging-in-Publication Data

Korea : managing the industrial transition.

(A World Bank country study)
Report written by D.M. Leipziger and others.
1. Industry and state--Korea (South)
2. Korea (South)--Commercial policy. 3. Monetary policy--Korea (South) 4. Korea (South)--Economic policy. I. Leipziger, Danny M. II. International Bank for Reconstruction and Development. III. Series.
HD3616.K853K67 1987 338.9519'5 87-2213
ISBN 0-8213-0887-4 (v. 1)
ISBN 0-8213-0893-9 (v. 2)

CURRENCY EQUIVALENTS

Currency unit	=	Won
US$1	=	W 880
W 100	=	US$0.11
W 1,000,000	=	US$1,136

GLOSSARY OF ABBREVIATIONS

AA List	-	Automatic Approval List
BOK	-	Bank of Korea
CCCN	-	Customs Cooperation Council Nomenclature
DMB	-	Deposit-Money Bank
EEC	-	European Economic Community
EPB	-	Economic Planning Board
ERP	-	Effective Rates of Protection
HCI	-	Heavy and Chemical Industry
KDB	-	Korea Development Bank
KDI	-	Korea Development Institute
KEPCO	-	Korea Electric Power Company
KEXIM	-	Korea Export Import
KHIC	-	Korea Heavy Industry Corporation
KLTCB	-	Korea Long-Term Credit Bank
MITI	-	Ministry of Trade and Industry (Japan)
MTI	-	Ministry of Trade and Industry (Korea)
NBFI	-	Nonbank Financial Institution
NHF	-	National Housing Fund
NIC	-	Newly Industrializing Country
NIF	-	National Investment Fund
OECD	-	Organization for Economic Cooperation and Development
OPEC	-	Organization of Petroleum Exporting Countries
POSCO	-	Pohang Iron and Steel Company
SMI	-	Small and Medium Industry
VCR	-	Video Cassette Recorder

THE CONDUCT OF INDUSTRIAL POLICY

Table of Contents

This Report was prepared on the basis of missions visiting Seoul in August and December, 1985 and was written by D.M. Leipziger (Mission Chief), Y.J. Cho, F. Iqbal, P. Petri, and S. Urata with contributions by S. Edwards (Consultant) and D. O'Connor and the assistance of S.Y. Song. Administrative assistance in Washington was provided by J. Tolbert.

Page No.

Page No.

List of Tables

List of Figures

List of Boxes

KOREA: MANAGING THE INDUSTRIAL TRANSITION

EXECUTIVE SUMMARY

Introduction

1. Korea is at an important juncture in its development. It has successfully weathered the structural problems of the 1979-81 period and has put its policy house in order. It has begun far-sighted programs of reform which it hopes will facilitate the industrial transition needed to push Korea towards the ranks of the industrialized countries. The economy is thus in a period of rapid change, and the conduct of public policy towards key sectors is undergoing a similar transformation.

2. This Economic Report therefore serves a number of purposes: first, it provides a review of recent economic developments and the principal issues influencing Korea's achievement of its Sixth Five-Year Plan goals; second, it addresses in a broad fashion the conduct of industrial policy; and third, it provides a selective yet in-depth look at certain key topics related to macroeconomic and industrial policy, including an examination of three industrial sub-sector case studies. With respect to industrial policy, the primary focus of the Report, the Korean incentive regime is assessed in light of the interventionist posture of Government in the 1970s and the ensuing reform movement, the process of trade and financial liberalization is analyzed, and Korea's industrial structure is discussed in some detail. Finally, the formulation of future industrial policy is addressed and issues are identified with respect to declining industries, conglomerate policy, and emerging industries.

3. The companion volume of this Report contains special topics pieces on Korea's industrial structure, the conduct of macroeconomic policy, the sequencing of trade and financial liberalization, the theory of industrial policy, and issues and evidence in industrial finance. It also contains three industrial case studies--on shipping, textiles and electronics--which were selected for didactic reasons to illuminate issues related to declining, transitional, and emerging industries.

Macroeconomic Overview

4. For the past several years Korea has followed a combination of tight monetary and fiscal policies and exchange rate depreciation in pursuit of three targets: price stability, current account balance and high GNP growth. While the use of contractionary aggregate demand policies cum depreciation runs the risk of a short-run, but potentially disruptive, recession, such a contingency did not materialize in Korea during 1982-84. On the contrary, investment and output grew while inflation and the current account deficit were substantially reduced, largely as a result of favorable external developments in oil and other commodities and in OECD growth. During 1985, investment and growth faltered, but the manner in which the economy was managed provided sufficient stimulus to prevent recession without sacrificing the hard won gains from previous conservative policies. Essentially, monetary restraint was moderately relaxed, fiscal expenditures were rearranged without

being increased, and exchange rate depreciation was accelerated. A domestic stimulus was thus provided, selectively targeted for investments in the export sector, a strategy that helped boost aggregate demand while maintaining the impetus towards external balance.

5. The job of demand management in recent years has been facilitated by the simultaneous implementation of supply-side policies. Trade and financial liberalization initiatives and industrial policy reform, undertaken to improve the efficiency of domestic markets, began to pay off in the form of higher domestic savings, a pass-through of low international inflation, a reduction of import dependence, and improvements in labor productivity and investment efficiency. While restrained macroeconomic policies have helped curtail inflation, part of inflation reduction has also come about through such supply-side developments. Thus, favorable external developments, such as the low rate of inflation among Korea's major trading partners, combined with a reduction in domestic trade barriers and an increase in the efficiency of import use, helped to keep downward pressure on domestic inflation. Similarly, while depreciation and a generally tight macroeconomic policy stance have improved the current account and reduced the rate of growth of external debt, part of this outcome is attributable to policies which have raised the domestic savings rate.

6. Prudent demand management policies combined with structural changes on the supply side not only enabled Korea to achieve its Fifth Plan (1982-86) targets, but also to be well positioned to take advantage of recent, very favorable events in the world economy including oil price declines, interest rate diminution, and enhanced OECD growth prospects. Indeed, inasmuch as 1986 should be a banner year for Korea, it should provide more leeway in macro policy formulation, and more importantly, the cushion needed to undertake some deferred policy reforms with respect to industrial and financial restructuring. It also provides a very favorable climate for trade liberalization to proceed on its preannounced course with a minimum of dislocation.

7. The aforementioned features of recent macroeconomic performance (discussed in detail in Chapter 2 of Volume II) provide a useful perspective from which to evaluate Korea's medium-run goals, prospects, and policy choices. These goals are expressed in the Sixth Plan to be implemented over 1987-91. The principal goals of the Sixth Plan are to achieve growth with price stability and a current account surplus. The latter objective is especially noteworthy because Korea aims to reduce its debt exposure relative to GNP. Gross external debt, which currently stands at about $45.6 billion, is targetted to increase by only $5 billion to $51.4 billion by the end of 1991. This is to be accomplished by achieving a surplus of domestic savings over investment (but not by reducing the investment rate) or equivalently, by achieving systematic surpluses on the current account by maintaining a high export growth rate of about 10%.

8. The prospects for export growth depend on a number of factors whose evolution cannot be predicted with precision. Perhaps the most important limiting factor will be the intensity of the protectionism that Korea may face in the future. Uncertainty regarding this makes it difficult to predict

Korean export growth, as is customary, on the basis of forecasts of OECD growth alone. On the positive side, trade and industrial policy reform should make it easier to diversify and expand exports, and sustained weakness in the price of oil, Korea's most important import, should make it easier for export, balance of payments and GNP targets to be achieved. In addition, the appreciation of the yen should enable Korea to increase exports in markets in which it competes with Japan.

9. As far as the prospects for further growth in the domestic savings rate (now at 28.6%) are concerned, the main considerations are that income growth prospects are bright and appropriate savings-augmenting policies (featuring high deposit rates and more flexible savings instruments) are in place. However, an environment of income growth and financial incentives to save is the minimum requirement for continued growth of savings. Given such an environment, the Sixth Plan target of a 33% savings rate by 1991 may be considered optimistic but reachable. The sector in which savings should probably be most encouraged is the corporate sector inasmuch as this would help to strengthen firm finances and bring corporate profitability objectives to the fore.

10. The importance of income growth for savings mobilization has a bearing on Government policy concerning external debt. If the strategy of debt reduction were to result in significantly lower growth, this could depress savings and possibly increase the investment-savings gap. Therefore, income growth should not be lightly sacrificed to achieve lower levels of debt exposure. Nor should additional debt be avoided at all costs, especially if the underlying payments imbalance which necessitates borrowing is transitory. Vulnerability to debt can be reduced without jeopardizing income growth by diversifying among sources of foreign capital, in particular by encouraging direct and portfolio foreign investment. Such investment could also play a valuable role in sustaining export growth in the future, partly by providing needed capital and technology and partly by relieving the pressure of protectionism. The larger the stake that OECD corporations have in Korea's industrial growth, the greater will be their aversion to protectionism in their home markets.

11. Korea has set ambitious targets for the remainder of the 1980s. A solid policy framework is in place to underpin these efforts and efficient and practical demand management techniques continue to bolster confidence that the Plan's objectives can be met. Korea's successful pursuit in recent years of the multiple goals of higher savings, higher exports and growth together with price stability should not blind it, however, to the inherent difficulties of such ambitious tasks. If the external environment is not as accommodating in the future as it has been in the recent past, if macroeconomic policy flexibility is constrained to an excessive degree by such features as the high leverage of Korea's corporations, or if supply-side initiatives are derailed, these difficulties and tradeoffs will become more readily apparent. In this context, Korea's future industrial policies will play a central role in determining whether or not these medium-term targets can be achieved.

Industrial Policy Overview

12. Korea is embarking on a major industrial transformation, which it hopes will bring it mature country status by the year 2000. Contrary to the last major shift in industrial structure, the heavy and chemical industry (HCI) push of the 1970s which established the capital base for some of Korea's recent export successes (such as shipbuilding, iron and steel, and now autos), this shift to technology-intensive products is being designed around essentially functional rather than industry-specific interventions. This is an important industrial policy change. At the same time, this policy, which places the primary emphasis on market incentives, is being complemented by explicit attempts to improve the efficiency of both factor and product markets. Hence the government's policy focus on both trade and financial liberalization.

13. Government's new industrial policy orientation is based on the premise that the incidence of market failures has been reduced in the process of development, and that direct intervention is no longer feasible or desirable in light of the economy's increasing sophistication. (See Chapter 4 of Volume II on the theory of industrial policy which governs intervention.) This realization stems in good measure from the HCI program which substituted bureaucratic judgment for market tests, was costly, and left scars on the economy in terms of distorted credit markets, overly-indebted firms, and a very high concentration of industrial power. Government therefore is aiming at limiting its interventions to classic public externalities (such as technology or human capital development) and actions to offset clear market imperfections. An example of the latter is Government's decision to reserve a share of domestic credit for small and medium industries (SMIs) to offset historical discrimination and prevent credit market domination by large firms. These measures should be seen as temporary interventions in the public interest, which will be curtailed once the efficiency of economic agents and the autonomy of financial institutions are sufficiently developed. Their existence, however, points to the importance of financial reforms.

14. In recent years, there have been a number of significant reforms in domestic financial markets, including the raising of interest rates, the privatization and incipient autonomy of commercial banks, and the creation of new financial instruments and markets. Many of the goals of further financial reform, including greater flexibility in interest rate determination, the development of stronger equity markets, greater banking autonomy, and more efficient sharing of risk between the corporate, financial, and public sectors will be important for the achievement of Korea's industrial objectives. Indeed the pace of financial reform may determine the speed with which Government is able to withdraw from industrial decision-making. This is especially true in the area of industrial restructuring which should, in mature markets, normally be handled by private creditors rather than Government. In this connection, one of the revised signals that creditors might be encouraged to send to the corporate sector is the importance of profitability (see Chapter 5 of Volume II). Without strong profit performance, further equity investment will be more difficult to attract, long-term capital and technology requirements will be harder to finance, and possible future market setbacks harder to sustain.

15. Domestic financial reform can be also seen as a prerequisite for the eventual opening of the capital account, although this should be approached cautiusly (as discussed in Chapter 3 of Volume II) and as a counterpart to trade liberalization. Korea has pursued trade reform vigorously in the continued belief that international prices provide the best price signals and are essential for increasing domestic competition and allocative efficiency. In pursuing that goal, Government has moved aggressively to shift an increasing number of import items to the unrestricted importation list and is at the same time implementing planned reductions in tariff levels. Government has used exceptional relief measures sparingly and has been resolute in following through on pre-announced import reforms, even in the face of disappointing overall economic conditions, such as in 1985. While further progress is still possible in increasing the import regime's transparency, Korea has come quite far in opening its markets to foreign competition.

16. There can be no doubt that the present course of trade liberalization will provide the economy with the proper set of incentives for efficient resource allocation. The course should be maintained, even in the face of the inevitable demands for relief by affected industries or potentially by those losing international competitiveness and wanting assured home markets. (The case study on textiles reviews the OECD experience in an industry facing declining comparative advantage; see Chapter 7 of Volume II.) Trade liberalization is, of course, a reciprocal activity, and there are clear risks that hard-won and farsighted reforms, implemented fairly rapidly, will come unraveled if Korea is subjected to either harsh protectionism abroad or excessive pressure to prematurely open sensitive markets. Much of the conflict in trade diplomacy in the 1980s has been caused by the changing environment; namely, the range of permissible conduct narrowed as trade volumes shrank, global growth lagged, and the mercantilistic policies of some trading nations caused ripple-effects for others. Recent sharp swings in exchange rates may well exacerbate large bilateral imbalances. Therefore, since protectionism poses the gravest threat to Korea's expansion, issues of trade diplomacy will require priority attention.

17. Korean planners see an industrial structure developing which places greater emphasis on technology and skill-intensive products (see Chapter 1 of Volume II). As a result, future Korean exports are seen to be competing with the advanced countries in areas such as electronics, autos, and semiconductors. To be successful, Korea will have to continue to push up its already high savings rate to finance the technology requirements without adding greatly to its external debt, be vigilant in keeping both the productivity-wage balance and the exchange rate on track, and assist the adjustments which will have to take place in terms of skill mix and employment shifts. Complementing these economic issues will be the necessity of successfully managing trade diplomacy, especially since Korea will be the new entrant in a number of sensitive export markets in the U.S. and will be directly competing with Japan, which is both a major creditor and a major intermediate input supplier. These issues serve to illustrate the high risk nature of Korea's emerging industry policy. (The case study on electronics provides further details; see Chapter 8 of Volume II.)

18. There are a number of areas in which the economy can be strengthened to both make it more resilient and increase its flexibility. Resiliency can be augmented by shoring up corporate finances and by reducing "moral hazard" behavior by which both borrowers and lenders undertake excessive private sector risk in the expectation that the public sector will ultimately shoulder losses. While this scenario has characterized recent public sector-led restructurings of the shipping and overseas construction industries, Government has signalled its new policy intent to avoid future industrial involvements as much as possible. (Recent government involvement in restructuring is examined in the case study on shipping; see Chapter 6 of Volume II.) Contrary to the preferentially financed huge capital investments of the HCI push, Government does not intend to be the risk partner for the country's emerging industries, such as electronics. Whether the public sector can achieve the degree of industrial nonintervention it desires and whether it can avoid being dragged into future industrial problems remains to be seen.

19. Based on the experiences of the OECD countries, where shifts in comparative advantage have often been resisted rather than accommodated, it may well prove more difficult for the government to remain aloof from declining industry problems than to establish industrial neutrality in the area of emerging industries. Korea's planned industrial transformation can be expected to produce more declining industry problems. Although some declines may be cyclical, others will be structural, and in those cases resources will have to be withdrawn from uncompetitive activities. This requires an effective "exit policy." At present, there is significant government involvement in restructuring at the industry-wide level due to overall financial circumstances; however, Government has not permitted these interventions to become a routine matter and has signalled to industry that small and large firms alike will be allowed to fail. Now that the signals are correct, further privatization of exit procedures will require strengthening the role of creditors, without which continued public sector involvement in future restructuring exercises unfortunately appears inevitable.

20. A related area of concern involves Korean conglomerates, the jaebol. Economic concentration has increased over time, and although the conglomerates pose perhaps the best chance for the desired technological leap of industry, they also embody large risks. It is important that these corporate risks (for example, in the advanced electronics field) not end up as public sector losses. Hence government efforts (largely unsuccessful to date) to strengthen jaebol finances and restrict them to main business lines as well as fiscal and financial incentives for SMI development. Once again, these measures are practical "second-best" policies, which substitute for the preferred market tests based on vigorous risk assessments and profit-maximizing activities of lenders. The long-term solution to the conglomerate issue is in large measure to strengthen the hand (and independence) of commercial banks and allow Government to be essentially neutral with respect to corporate size. Domestic competition should of course be maintained, most easily through the continued pursuit of import liberalization, or anti-trust measures in the case of domestic goods.

21. With respect to emerging industries, the era of explicit import substitution or directly promoted infant industries is largely past, and Korea

has opted instead to use functional incentives to foster growth in new industries at a pace which is socially desirable. That is not to say that government will have no role in guiding the economy's future industrial development, but rather that it will concentrate on providing efficient industrial signals, support for R&D and technology development, and information in cases where markets are short-sighted or otherwise imperfect. Government's recently announced policies are incorporated in the new Industrial Development Law. While the legislation does not include any "MITI-style" guidance mechanism, it does provide for public-private sector councils to reach industrial consensus. Recent pronouncements have indicated that Government does have an industrial vision; however, it remains to be seen how that vision is actually effectuated.

22. Many reforms--including import liberalization, financial adjustments, and revised industrial policies--can best be implemented in a favorable economic climate. By all accounts, 1986 should provide such a salutary environment, and that opportunity should not be missed. Recent prospects point to a vigorous expansion, with low inflation and strong balance of payments performance. At the firm level, generally lower input costs and a favorable exchange rate may boost profitability, and provide an opportunity for the corporate sector to shore up its finances. This should be encouraged. To the extent that sharp oil price declines and falling interest rates provide the economy with "windfall gains", the timing may be propitious to deal with the issue of non-performing loans of troubled industries and other structural problems; for example, the cushion provided by recent economic events might enable Government to deal surgically with some of the debilitating problems in the financial sector rather allow them to slowly drain public resources and continue to inhibit the sector's efficiency.

Principal Conclusions

23. The Report's major conclusions are that Korea has managed to be pragmatic and flexible in the conduct of both its macroeconomic and industrial policies, that even in the presence of economic biases it has been careful to integrate the components of its incentive policies to be at least modestly pro-export, and that it has tried as much as possible not to move contrary to market signals. There have clearly been lapses in policy, although even during the HCI episode, which proved costly, exports remained the ultimate objective which forced industry to be competitive at world prices. Now that Korea is moving into perhaps the final phase of its industrial transformation--having successfully moved from an agricultural economy to a manufacturing economy based on productive labor, and partially to a capital-intensive exporter--it faces new challenges. The move to higher technology and skill output will require a continued favorable climate of economic management supplemented by ever-more efficient markets and continually less direct intervention in those markets. In earlier versions of this Report, the Bank has offered some policy views concerning trade and financial reform, macroeconomic policy tradeoffs, and the broad conduct of industrial policy.

24. Policymakers have embarked on an ambitious, yet carefully balanced course of industrial transformation, and they are clearly aware of the challenges they face. Korea's proven past performance, its resolve and pragmatism, and its economic drive lead to the expectation that these efforts will not go unrewarded.

CHAPTER 1: MACROECONOMIC DEVELOPMENTS: REVIEW AND OUTLOOK

A. Recent Macroeconomic Experience (1982-85)

Basic Developments

1.01 Korea's recent macroeconomic history is essentially a chronicle of the Fifth Five Year Plan which was initiated in 1982, was revised and relaunched in 1984, and is to be concluded in 1986. The Fifth Plan was formulated in a period of great stress for Korea. A series of adverse events rocked the economy in 1979-81. Among these were sharp increases in the prices of oil and interest rates, and the onset of an international recession.[1] As a consequence of these shocks, Korea experienced the first-ever decline in GNP growth (by 5.2% in 1980) together with high rates of inflation (29% in 1980 and 16% in 1981) and large current account deficits (9% of GNP in 1980 and 7% in 1981). The savings rate tumbled from 28% in 1979 to 22% in 1980, and external debt ballooned from about $15 billion in 1978 to about $32.5 billion by the end of 1981.

1.02 Given this background, the Plan correctly placed great emphasis on a program of stabilization and adjustment, featuring conservative monetary and fiscal policies, exchange rate depreciation, and initiatives to reduce the rate of growth of wages and the intensity of energy use. Despite conservative policy intentions, however, the assumptions of the Plan were overly pessimistic. It was assumed, for example, that the prices of oil and other raw materials would continue to rise and would lead to a domestic inflation rate of about 10% and a current account deficit of about $4 billion per annum during the Plan period.

1.03 During the first two years (1982 and 1983), remarkable progress was made in reducing inflation as well as the current account deficit. By the end of 1983, for example, domestic inflation stood at 4% as compared to the Plan projection of 11%, while the current account deficit was $1.6 billion as compared to a projection of $4.4 billion. The rate of growth of GNP, meanwhile, averaged 8.5% during 1982-83, a point higher than projected in the Plan. So rapidly had progress been made and so much had the circumstances changed that Government drew up a Revised Plan to cover the remaining three years (1984-86), the chief features of which were much more optimistic projections of inflation (under 2%) and the current account deficit (in surplus in 1986) together with an unchanged projection for GNP growth of 7.5% per annum and a continuation of the overall conservative policy stance.

1.04 It is useful to note parenthetically that Plan targets were not met in two important respects. Commodity exports were only $23 billion in 1983 as compared to a target of $30.5 billion and the domestic savings rate reached only 24.4% rather than the 25.7% targetted. The reduction in the current

---

1/ In addition to these external shocks, two domestic developments, poor harvests and the assassination of President Park, added to Korea's woes.

account deficit and, inter alia, in the need for external financing, was due largely to the substantial shortfall in imports which reached a level of only $25 billion as compared to a projection of $34 billion. The domestic investment rate, meanwhile, was only 27.6% in 1983, much below the projected level of 31.1%. As a consequence, when the Fifth Plan was revised both export and investment targets were adjusted downwards. However, the expected rate of growth for exports was not substantially changed.

1.05 The performance of the economy in 1984 was in keeping with that of 1982 and 1983. The inflation rate remained low (at 4%) and the current account deficit declined further (to 1.7% of GNP), while the economy grew at 8.4%. During 1985, however, the pace of growth faltered. A sharp fall-off in export growth led to a reduced rate of GNP growth (5.1%). In other respects, however, progress continued to be made. Prices remained relatively stable, as the inflation rate did not rise above 4%, and the current deficit continued to decline (to 1% of GNP). However, exports were virtually unchanged from the 1984 level of $30 billion and considerably below the expected level of $33 billion. Once again it was a sharper drop in imports (relative to projections) that brought about the favorable current account result.

1.06 Despite disappointing performance in 1985, it is possible at this juncture to predict that Korea will meet its major Plan targets by the end of 1986. An extremely favorable sequence of developments in the first quarter of 1986 has made it likely that the shortfalls experienced in 1985 will be adequately made up in 1986. Three principal developments are expected to help: (a) the recent steep decline in the price of oil; (b) the decline in international interest rates; and (c) the recent sharp appreciation of the Japanese yen. All of these developments should make it easier for Korea to achieve a balance of payments surplus and a high GNP growth rate without exceeding its targets for external debt acquisition and inflation. The decline in the price of oil should help Korea directly by reducing its import bill and indirectly by sparking a higher rate of world economic growth and trade. The decline in interest rates should also improve Korea's balance of payments by reducing debt service payments. These two developments should also make it easier to maintain price stability. Finally, the appreciation of the yen should enhance Korea's export prospects, especially in large markets (such as the US and EEC) in which Korea competes actively with Japan in such important product lines as automobiles, electronic goods and iron and steel products.

The Policy Environment

1.07 The macroeconomic policies followed to achieve Fifth Plan targets were essentially tight money, nearly balanced budgets and exchange rate depreciation--the standard prescription for a small open economy in search of macroeconomic balance in the aftermath of high inflation and a large external deficit. Monetary policy featured a sharp deceleration in the rate of growth of net domestic assets from 39% in 1981 to 14% in 1984. Nominal bank lending rates, which are controlled by the government, were permitted to rise during 1983-85 after having been reduced sharply in 1982 in the wake of a sharp fall in inflation. As a consequence, real interest rates have continued to rise and have in fact doubled (from 4% to 8%) during the last five years.[2/]

1.08 Fiscal policy has been characterized by a declining ratio of budget deficits to the GNP, from 4.4% in 1982 to 1.5% in 1985. Moreover, the ratio of expenditures to GNP was brought down from about 24% in 1982 to 21.3% in 1983, a level at which it has stayed since. Thus, budget deficits have been reduced not by raising taxes, but by restraining expenditures.

1.09 As far as exchange rate management is concerned, the nominal exchange rate was depreciated in small steps up to mid 1984 after a major devaluation of 24% in early 1980. Since the dollar is prominent in the currency basket to which the won is pegged, its appreciation during 1984 also caused an appreciation of the won. This compounded the difficulties brought on for Korea's exports as a result of the stagnation of the OECD economies in 1985. To revive exports, Government depreciated the won sharply in 1985 such that its nominal value in terms of the dollar dropped from 810 to 890. The real effective exchange rate has depreciated from an index level of 100 in 1982 to 82 by the end of 1985.

1.10 Clearly aggregate demand management since 1981 has been contractionary when compared to the policy environment of 1980-81. Such a policy stance is normally expected to reduce inflation and the current account deficit, but runs the risk of a short-run, but potentially disruptive, recession. Moreover, exchange rate depreciation not only heightens this risk, but also raises the risk of inflation in the short run (viz., the so-called contractionary devaluation problem). A remarkable feature of Korea's recent macroeconomic history, however, is that such a contingency did not materialize during 1982-84. On the contrary, both investment and output grew while inflation and the current account deficit were substantially reduced. Investment accelerated to an average of 18% during 1983 and 1984. Similarly, GNP grew at 5.6% in 1982 and accelerated to an average of 10% during 1983 and 1984 (See Table 1.1).

Strategic Factors

1.11 Closer scrutiny suggests that external developments and supply side initiatives have been additional important determinants of recent macroeconomic performance. Furthermore, domestic demand management has been conditioned by a structural change in the financial sector, namely the growth

---

2/ During 1985, the slowdown in exports and the slackening of GNP growth prompted some monetary relaxation, and net domestic assets were permitted to expand at 24%. However, this was clearly an exception to the generally restrained stance that characterized monetary policy under the Fifth Plan. The target rate of growth of net domestic assets in 1986 is 14%, although care must be exercised as noted in para. 1.14 to consider the rate of monetary expansion in the NBFIs.

Table 1.1: RECENT MACROECONOMIC DEVELOPMENTS
(%)

	1982	1983	1984	1985
Policy Stance				
Monetary				
NDA growth /a	34	18	14	24
Interest rate:				
Nominal /b	12	10	11	12
Real /c	5	7	7	8
Fiscal				
Deficit/GNP	-4.4	-1.6	-1.4	-1.5
Expenditures/GNP	24.2	21.3	21.1	21.4
Exchange rate				
REER change /d	0.3	-7.3	0.2	-11.0
Macroeconomic Performance /e				
GNP growth	5.4	11.9	8.4	5.1
GNP deflator	6.6	3.9	3.8	3.6
Current account/GNP	-3.8	-2.2	-1.7	-1.0
Investment growth	0.1	17.5	18.6	1.6
Savings rate	20.9	25.3	27.9	28.4
Investment rate	28.6	29.9	31.9	31.2

/a NDA refers to Net Domestic Assets of banking system.
/b Ceiling lending rate on loans of one-year maturity by commercial banks.
/c Nominal rate adjusted by GNP deflator measure of inflation.
/d Rate of change in the value of the real effective exchange rate (REER).
/e Data reflect the recently adopted new System of National Accounts methodology.

Sources: Bank of Korea, Economic Statistic Yearbook. Bank of Korea, New System of National Accounts.

of nonbank financial institutions (NBFI's), a new financial market segment that was actively promoted by reforms undertaken since 1980 (as described in Chapter 4, Volume I in some detail).

1.12 The importance of external influences is to be expected in a trade-dependent economy such as Korea.[3/] During 1982-84, favorable external

---

3/ The following statistics reveal the importance of trade to Korea. Exports account for about 40% of GNP. The elasticity of Korean exports to OECD income is very high, ranging from 3.5 to 4.5. Petroleum accounts for over a fifth of imports and mineral products, principally oil, have a 33% weight in Korea's import price index.

developments, such as the decline in the prices of oil and other commodities and the economic recovery in the OECD countries (and especially in the US), enabled Korea to reduce inflation as well as to maintain a high rate of investment and growth despite relatively contractionary macroeconomic policy. External developments were also important to Korea's economic performance in 1985. As the US and other OECD economies entered a period of stagnation from mid-1984 onwards, Korean exports suffered and both investment and growth faltered.

1.13 The job of demand management was made easier by the simultaneous implementation of supply side policies. Trade and financial liberalization initiatives and industrial policy reforms undertaken to improve the efficiency of domestic markets began to pay off in the form of higher domestic savings, a pass-through of low international inflation, reduced import dependence, and improvements in labor productivity and investment efficiency. While restrained macroeconomic policies helped curtail inflation, part of inflation containment has come about through supply side, competition-enhancing steps such as import liberalization. Thus, favorable external developments, such as the low rate of inflation among Korea's major trading partners, combined with lower trade barriers and an increase in the efficiency of import use to keep downward pressure on domestic inflation. In addition, wage control policies, such as mandated reductions in the rate of growth of public sector wages and moral suasion in the case of private sector wage settlements, helped reduce nominal wage growth from about 21% in 1980-81 to an average of 8.5% during 1984-85.

1.14 The relative size of the nonbank financial institutions (NBFIs) reached a significant enough level in the 1980s to affect the conduct of monetary policy.[4/] In essence, this development has increased the uncertainty regarding the ultimate effect of monetary policy, exercised through the primary banking sector, on key macroeconomic variables inasmuch as monetary developments in the NBFI sector can offset developments in the formal credit market. During 1983-84, for example, the authorities sharply reduced the rate of growth of domestic credit through banks, but this action was partially offset by developments in the NBFI sector where the rate of growth of credit did not decline as sharply (and in fact rose during 1984) and real interest rates generally fell. Similarly, when a more relaxed monetary stance was adopted in 1985, it was offset to some extent by contraction in the NBFI

4/ The non-bank financial institutions consist essentially of insurance, securities and short-term finance and investment companies. They are permitted to raise cash and lend at rates higher than those permitted regular banks and face fewer restrictions in their credit allocation policies. They are also allowed to issue as well as guarantee bonds and commercial paper, the latter at market determined interest rates. As a consequence of these regulatory advantages, NBFIs have grown significantly in recent years, partly at the expense of the regular commercial banks. Their share of deposits in the financial system grew from 27% in 1980 to 42% in 1984, while that of regular banks declined from 73% to 58% (see Table 2.5, Volume II).

segment. As a consequence, broader measures of liquidity which include the deposit and credit creating activity of NBFI's should also be considered in the process of macroeconomic performance determination. However, these measures are both less amenable to government control and less predictable in the absence of a sufficient history of policy experience.

1.15 The principal lessons of Fifth Plan experience have been that external and supply side developments have been as important in determining macroeconomic performance as domestic aggregate demand policy. In addition, structural developments in the financial sector have rendered monetary control more difficult and its effects more uncertain. These factors will undoubtedly play an important role in the experience of the Sixth Five-Year Plan which is to be implemented over 1987-91.

B. Sixth Plan Goals

1.16 Perhaps the most prominent goal of the Sixth Plan is the achievement of a reduction in Korea's net external debt. This is targeted to decline from its end-1985 level of $34.2 billion to $29.5 billion by end-1991. It is to be achieved by means of a sharp reduction in the rate of growth of gross external debt and a sharp increase in that of Korean-held foreign assets. The outstanding gross external debt, which currently stands at about $45.6 billion, is targeted to increase by only $5 billion over the next 6 years (to $51.4 billion) at average annual increments of less than $1 billion. Korean-held foreign assets such as international reserves and export credits are expected to increase from about $11 billion in 1985 to about $22 billion in 1991. Associated with the net and gross foreign debt targets are the facilitating targets of achieving a surplus of domestic savings over domestic investment or equivalently, the achievement of surpluses on the current account.

1.17 These facilitating targets are, however, to be achieved without jeopardizing the target of a rate of growth of 7%. Hence the two gaps are to be closed and reversed not by reducing investment and imports but by significantly increasing savings and exports. Gross national savings are targeted to rise from 28.6% in 1985 to 33% by the end of 1991, while exports are targeted to rise at an average rate of 10% (in real terms) during 1987-91. Thus, during the Sixth Plan, Korea aims to effect a net transfer of resources out of the country in order to improve its external debt exposure (see Table 1.2).

1.18 The Sixth Plan envisages a policy environment similar to the one that has characterized the implementation of the (revised) Fifth Plan. Specifically, trade and financial liberalization are to be continued. On the trade side, almost 95% of imports are to be allowed entry without quota restrictions (and with declining tariff protection) by 1991. On the financial side, increasing autonomy is to be permitted to commercial banks in their credit pricing and allocation decisions and the financial sector is to be "internationalized" by progressively opening it up to foreign banks and securities firms. Special measures are to be taken to attract direct and portfolio foreign investment. In addition to trade and financial liberalization, the macroeconomic policy regime is to continue to be characterized by prudent and flexible fiscal, monetary and exchange rate policies. Budget deficits are to remain under 2% of GNP and fiscal policy is to have a greater social orientation. Monetary growth is to be kept in line with the growth of GNP and is to

Table 1.2: SELECTED MACROECONOMIC TARGETS

	Units /a	1985	1986	Sixth Plan Targets /b					1987-91 (annual increase rate)
				1987	1988	1989	1990	1991	
GNP (current prices)	$ bln	83.1	91.0	102.4	115.2	129.1	144.0	166.6	12.0
GNP (1980 prices)	$ bln	86.5	93.4	100.4	107.9	115.5	123.0	131.0	7.0
Growth rate	%	5.1	8.0	7.5	7.5	7.0	6.5	6.5	7.0
Total investment to GNP	%	31.2	31.3	31.4	31.5	31.6	31.6	31.5	31.5
Domestic savings to GNP	%	28.4	30.1	30.9	31.5	32.1	32.6	33.0	32.0
Foreign savings to GNP	%	3.1	1.2	0.5	0.0	-0.5	-1.0	-1.5	-0.5
GNP deflator	increase rate (%)	3.6	2.5	3.5	3.5	3.5	3.5	3.5	3.5
Wholesale prices		0.9	-2.0	2.0	2.0	2.0	2.0	2.0	2.0
Current account balance	$ bln	-0.8	-0.5	1.0	1.4	1.7	2.4	3.0	
Trade Balance	$ bln	-0.0	1.6	2.4	2.7	3.1	3.7	4.3	
Exports of commodities	$ bln	26.4	30.8	34.4	38.4	43.0	47.8	53.1	11.5
Imports of commodities	$ bln	26.4	29.2	32.0	35.7	39.9	44.1	48.8	10.8
Invisible trade balance	$ bln	-0.8	-1.1	-1.4	-1.3	-1.4	-1.3	-1.3	
Total external debt	$ bln	46.8	48.1	49.4	50.7	51.7	52.1	51.8	
Total external assets	$ bln	11.3	11.8	13.6	15.8	18.0	20.2	22.4	
Net external debt	$ bln	35.5	36.3	35.8	34.9	33.7	31.9	29.4	

/a Based on new System of National Accounts. For a comparison of the old and the new SNA methods, see the Statistical Appendix. By and large, the new SNA by including previously unrecorded or underrecorded economic activities yields higher GNP growth estimates.

/b These are preliminary estimates based on data available as of July 1986.

Source: Economic Planning Board.

be keyed to financial liberalization. The exchange rate is to be managed competitively in pursuit of external surpluses. Several efficiency-enhancing initiatives are to be taken in the area of industrial policy; in general the initiatives are characterized by a shift from sectoral incentives to functional incentives and from a preference for large scale enterprises to small and medium industries.

1.19 The following sections examine the feasibility of the Sixth Plan targets. Likely developments in export performance, the rate of domestic savings and access to external finance are assessed. The appropriate policy responses to future shocks are discussed in the light of recent macroeconomic policy experience and principal risks are noted, in particular those associated with domestic policy offsets to external events.

C. The Outlook for Exports

1.20 Korea's medium-run plans are predicated on the assumptions that OECD economic growth will average 2.7%, OECD import demand will grow at an average of 5%, and Korean exports will grow at 10% (in real terms) during the next five or six years. These are critical assumptions for the simple reasons that exports are the single most important determinant of growth in Korea and the elasticity of exports to OECD income growth is very high. Thus, if this set of assumptions is not validated by actual experience, Korea's growth performance will falter and its expected average growth rate of 7.0% will not materialize.

1.21 Recent projections of world and OECD economic growth rates indicate that while the slowdown evidenced in 1985 will carry over into 1986 the average growth rate over the next five years will probably be close to the rate on which Korea has based its Sixth Plan. There is a tendency to be preoccupied with the relatively lackluster global performance in 1985 and to generalize from it. This risks confusing cyclical behavior with long-term trends. While it is always hazardous to predict the length of a cycle stage it is generally anticipated that an upturn should begin in 1987. The recent sharp drop in the price of oil may produce an upturn even earlier and of greater strength than expected in forecasts made only a few months ago. The probability of OECD income growth falling below 2.7% is low (see Table 1.3). What cannot be taken for granted, however, is that the rise in OECD incomes will generate the same sort of demand for Korean exports in the future as has been the case up to now. Intensifying protectionism abroad, decreasing Korean competitiveness in certain products and increasing competition in others are reasons why Korea cannot afford to be complacent.[5/]

5/ A rule of thumb that has characterized Korea's relationship to foreign markets is that Korean exports grow at three times the rate of growth of imports in its major trading partners (in volume terms). This rule of thumb was broken in 1984 and 1985 when the ratio of Korea's export volume growth to foreign import growth fell below unity. This may be a sign of increasing difficulties in export markets.

Table 1.3: WORLD TRADE AND GROWTH
(%)

	(1970-79) average	1982-86	1987-91
GNP Growth			
OECD average	3.3	2.5	2.7
United States	3.1	2.7	3.0
Japan	4.9	4.0	3.9
Europe	3.3	1.4	2.4
Trade Volume			
World average	5.4	3.4	4.7
United States	5.7	7.5	4.3
Japan	7.3	4.8	7.2
Korea (exports)	23.2	8.9	10.0

Sources: Wharton Econometrics; Data Resources Inc.; OECD Economic Outlook.

1.22 Two important factors that permit a more optimistic assessment of Korea's export prospects for the remainder of the decade are the recent sharp decline in the price of oil and the appreciation of the yen relative to the won. The decline in the price of oil has both direct and indirect effects. The direct effect will occur through the reduction of the oil import bill: approximately $200 million ought to be shaved from the import bill by every $1 drop in the price of oil. The indirect effect will occur through the boost given to OECD growth which should, in turn, provide a fillip to Korean exports. This effect should be assessed, however, in the light of the fact that the main OECD beneficiaries of the oil price decline will be Japan and some EEC countries and not the US. It is unlikely that growth in the EEC will lead to significantly higher volumes of imports from East Asia, since labor market rigidities and union power remain prevalent there. Further growth in Japan may be of greater help to Korean exports especially if it is combined, as at present, with an appreciating yen. Nevertheless, the difficulties of penetrating the Japanese market are not to be underestimated. The appreciation of the yen should be of more benefit to Korea in third markets where it competes with Japanese exports.

Protectionism

1.23 Evidence of significant protectionism against Korean exports in OECD countries is provided in Table 1.4. In the last 5 years, between 35% and 43% of total Korean exports have been under one form of restriction or another. For some products, such as footwear, textile and silk products, restrictions have typically covered over two thirds of total exports. It is also the case that the scope of protectionism has been widening. While only light industrial products were covered in the seventies, restrictions now extend to heavy industrial item such as steel products and high technology items such as television sets and other electronic products. As Korea has diversified partly to get around the protectionist wall, the wall itself has been extended. While quota arrangements for some exports have redounded to Korea's advantage in that a stable market share has been assured and higher profit margins have been earned, the overall consequences of protectionism have been detrimental to Korea's growth in the past and may be the most significant impediment to growth in the future.

Table 1.4: KOREA: TREND OF RESTRICTIONS ON MAJOR EXPORTS TO THE OECD, 1981-85 (%) /a

	1981 /b	1982	1983	1984	1985
Textile products	44.6	55.8	59.1	64.3	64.6
Steel products	30.3	43.0	39.0	47.3	27.1
Footwear	69.8	78.4	79.5	85.2	87.5
Fishery products	39.9	26.6	26.3	28.8	23.5
Electronic products and parts	4.8	16.2	16.3	12.2	13.4
Silk products	79.8	79.3	71.9	67.6	66.7
Cutlery	29.1	30.7	79.6	72.7	24.2
Tires and tubes	4.6	5.2	7.0	50.0	3.6
Television sets	...	28.3	51.2	40.8	35.7
Total selected products	35.2	42.8	39.8	41.2	36.6
Memorandum item: Share of selected products in total exports	(...)	(27.5)	(26.5)	(28.5)	(25.3)

/a Value of Korean exports under restrictions to total exports of the selected products.
/b Break in series.

Source: Data provided by the Korean authorities.

1.24 The appeal of protectionism is powerful because of the special characteristics of the industries most threatened by import competition. Textiles, footwear and steel, for example, are all industries which are characterized in the OECD countries by highly visible and vocal industry and labor interests. The industries are geographically concentrated and tend to be by far the most important providers of jobs in their areas of location. This aspect enables the industries to bring great pressure to bear on political representatives and thereby to maintain the impetus for trade restrictions regardless of the consequences for the national interest. These industries are too big and too important, both politically and economically, to be easily abandoned to overseas producers. Korea must plan its strategy with this limitation in mind. Recovery and growth in the OECD economies through the rest of the 1980s may weaken the demand for protection somewhat but, given the nature of the labor markets and political interests involved, it is unlikely that the protectionist curtain will lift enough to enable Korea and other NICs to increase substantially their share in OECD markets.

1.25 A specific analysis of the US market is instructive. Compared to Japan and the EEC, the US remains relatively open to Korean exports. While it has powerful protectionist lobbies (in textiles, for example), they are not as powerful as those of the EEC where labor unions are more entrenched and labor markets more rigid. In some cases, EEC governments themselves have investment stakes in industries affected by Korean competition and thus a natural partnership exists between Government and labor in those instances. Similarly, while the US has some tariff and nontariff restrictions on imports, its system is not as comprehensive or impenetrable as the Japanese system. Despite its relative openness, however, the US market should not be taken for granted by Korean exporters.

1.26 The large US trade deficits run in 1984 and 1985 have created hostility to free trade and an unfortunate interest in favor of "fair" trade, a euphemism essentially for managed trade. The trade deficit is now increasingly perceived both in overall terms and in bilateral terms. In overall terms, the US might be less concerned about accepting more imports from Korea, provided, for example, that Japan either accept reductions in its share of the US import market or open its own markets further to US exports. At the same time, active negotiations are under way to deal with bilateral imbalances and ways are being sought to compel countries such as Korea to open up their markets to US goods and services. In a milieu of managed trade, it is not necessarily the country with the lowest production costs that garners the greater share of the available gains. Instead, the ability to engage in trade diplomacy may be more important. The net result, of course, of any effort to manage trade would be reduced potential living standards worldwide.

1.27 A look at Korea's bilateral trade balances (see Table 1.5) indicates that while Korea has, on average, run surpluses with the US in recent years, it has run large deficits with Japan. In 1985, it ran a $4.3 billion surplus with the US and a $3.0 billion deficit wtih Japan. If deficits with Japan could be reduced, Korea could attain export targets consistent with its GNP growth targets without putting excessive pressure on its trade relations with the US. It would also then have less incentive to engage in import substitution activities which are at present driven by the desire to keep imports from Japan at a tolerable level. Thus, the future of Korea's trade relationship with the US and the scope for trade liberalization are intrinsically linked to its trade with Japan. Although until now, Korea has had limited success in penetrating Japan's domestic market, there may be greater scope for export gains in the next year or two if the recent appreciation of the yen is sustained. In the short run, the yen appreciation appears to have worsened the trade deficit with Japan, consistent with normal "J-curve" effects.

Table 1.5: KOREA'S BILATERAL TRADE BALANCES
($ billion)

	1975	1980	1981	1982	1983	1984	1985
US	-0.3	-0.3	-0.4	0.3	2.0	3.6	4.3
Japan	-0.1	-2.8	-2.9	-1.9	-2.8	-3.0	-3.0
Europe /a	0.2	1.0	0.9	1.6	1.1	0.5	0.3
Other	-2.0	-2.7	-2.5	-2.4	-2.0	-2.5	-2.5
Total	-2.2	-4.8	-4.9	-2.4	-1.7	-1.4	-0.9

/a Belgium, France, Germany, Italy, Netherlands, Norway, Sweden and UK.

Source: Bank of Korea, Economic Statistics Yearbook.

Changing Comparative Advantage

1.28 Another important factor in the export scenario is Korea's declining comparative advantage in some traditional exports. Textiles are a good example. These have long been the single most important foreign exchange earner and provider of employment. The sector accounts for a quarter of export earnings and of manufacturing employment. In recent years, however, the rate of growth of textile exports has slipped far below those experienced in the 1970s. While OECD restrictions embodied in the Multi-Fiber Arrangement (MFA) have been partly responsible, a major cause of weaker performance has been increasing competition from other Asian textile producers whose labor costs are below Korea's and whose technology is fairly similar.[6/] While recent macroeconomic policies have helped improve Korea's competitiveness in some light industry products (via depreciation and wage restraint), the fact remains that a structural adjustment away from products that have depended largely on low labor costs is called for. This will be hard because of its short-run implications for employment; however, there may be few alternatives. Devaluation would be an inappropriate tool to help specific export

---

6/ For example, earnings per hour in Korea's textile sector rose sharply from 78¢ in 1980 to $1.54 in 1982. This placed it above Hong Kong ($1.40) and Korea's other major competitor ($1.43) in higher-grade textiles, and far above such countries as Pakistan ($0.37) and India ($0.66) among its competitors in low grade products. Korea is feeling the same sort of pressure from such countries in textiles that it applied to the OECD countries in an earlier period. Trade competition from China has become especially fierce. China's share of the US market for imported apparel, for example, rose from 1% to 8% during 1978-82, while Korea's share stayed constant at 17%. For details on textiles, see the case study in Volume II of this Report.

industries since it affects import costs and export competitiveness across the board. Moreover, if rising wage and raw material costs are the principal cause of the decline in competitiveness, the appropriate response is an industrial policy which eases the exit of firms and workers out of the particular industry (see Chapter 5).

1.29 Some Korean exports are also likely to suffer, over the medium term, from worldwide overcapacity in such industries as shipbuilding, shipping and overseas construction, which are all of considerable importance to Korea's export earnings. While a cyclical upturn in the world economy will improve prospects in other areas, it may not provide the same magnitude of stimulus to exports and growth in the next five years that it did in the 1970s, due to changes in Korea's export structure and in its markets. It should be pointed out, for instance, that the geographical diversification of Korean exports has narrowed in recent years, such that the US alone took in about 38% of the total in 1985 (compared to 32% in 1982).

1.30 Given the dimmer demand prospects for traditional export items, given the competition from lower cost countries in traditional light industries, and the overall prospect of protectionism in OECD markets, there is ample reason to expect Korean exports to grow at a lower rate than in the past.[7] Korea has demonstrated remarkable flexibility in developing new export lines, however, and successful export diversification is likely to continue. New product areas that hold considerable promise are automobiles and automotive parts and sophisticated electronic products such as semiconductors, personal computers and similar high-tech items. High rates of export growth have been achieved in these areas. Whereas Korea's export drive in the 1960s and 1970s was propelled by wage-competitive, unskilled labor, the next wave of exports will be similarly built on skilled labor. A supporting infrastructure of R&D, technical education and training is already well established and further investments are envisaged.[8]

7/ The share of such items as textiles, clothing, footwear, iron and steel products, ships and other light and heavy industrial products that face tighter and tighter export markets is well over 60% of Korea's present exports.

8/ While promising starts have been made, the downside risks to forging ahead in the automobile and semiconductor sectors should also be taken into account. Both sectors are risky for Korea in that they are characterized by the presence of powerful oligopolistic participants. This is an additional source of the impetus for trade management and orderly market sharing agreements which make it especially difficult for new entrants to obtain footholds. Korea will eventually have to reach some sort of automobile and semiconductor trade agreement with the US and Japan and should not expect to make substantial inroads in either market simply on the basis of lower production costs. The market-sharing agreements eventually reached may or may not be compatible with large investments in capacity in these products. Prior to reaching a new sustainable trade pattern, there will be inevitable friction as export shares soar and/or prices fall on the basis of newly emerging scale economies.

Assessment

1.31 In sum, the feasibility of the Sixth Plan export growth target (of 10% p.a. on average) depends on a number of factors whose evolution cannot be predicted with any precision. Perhaps the most important limiting factor will be the intensity of the protectionism that Korea faces in the future. Uncertainty regarding this makes it difficult to use OECD output growth forecasts as a reliable guide to the effective demand for Korean exports. On the positive side, trade and industrial policies that make it easier to diversify exports, and sustained weakness in the price of oil, an important determinant of world output and trade growth, will make it easier for export growth (and GNP growth) targets to be achieved.

1.32 The prospects for export growth must be distinguished from those for improvements in the current account. It is worth noting that while Korea has generally been overly optimistic with respect to export growth targets in recent years, it has nevertheless been successful in reducing the current account deficit sharply. This has been due to favorable developments on the imports side, viz., declining oil prices and reduced import-dependence measured in terms of the imports necessary to produce exports, investment goods, and final demand (see Chapter 1, Volume II). It is likely that the oil price and interest rate developments of recent months will help Korea achieve a surplus in its current account in 1986 and beyond, despite protectionism and other difficulties on the export side and despite the fact that imports from Japan, the second most important source of imports for Korea, will be considerably more expensive as a result of the appreciation of the yen. The oil price windfall is so large that it should dominate all offsetting effects in the short run.[9/] In the longer run, the behavior of the current account will also be influenced by the income effects of the oil price drop, which should lead to higher imports, and by the structural developments on the exports side that have been described in some detail in this section.

D. The Savings Rate

1.33 The gross national savings rate (GNSR) has been characterized by an upward trend since the early 1960s broken on a few occasions by sharp spurts and reversals. While a number of economic, demographic and cultural factors have undoubtedly shaped this trend, the dominant influence has been that of rapid income growth. This is clearly shown by econometric analyses of savings behavior in Korea which, in common with such studies for other countries, uniformly show the highest elasticities and greatest explanatory power for

---

9/ During 1985, Korea purchased approximately 200 million barrels of oil at an average price of about $27 per barrel. If the average price of oil stabilizes at $15 and Korean purchases remain constant at the 1985 level, the windfall gain in import-savings will be $2.4 billion. To this should be added about $300 million from each percentage point drop in international interest rates.

variables denoting income level and growth.[10/] The rapid rise in the GNSR during 1975-79 was associated with the rapid rise in income in that high inflation, high growth period. The subsequent nosedive in the GNSR (from 27.8% in 1979 to 21.3% in 1980) was the result of the sharp oil price-induced decline in GNP growth (-5.2%) in 1980. Since then, as income growth has recovered so have savings.

Components of Savings

1.34 Disaggregation indicates that while household savings have contributed the most to the increase in the GNSR since 1980, they are also the most variable source of savings (see Table 1.6). Incomes per capita (and per household) have risen tremendously over the past two decades and have provided households with both a wider range of aspirations for which to save and the necessary margin of above-subsistence income which to channel into savings and thereby convert aspirations into future possibilities. Korean experience has been very similar to that of Japan in this regard. The two countries share similar conditions which have provided a spur to household savings: among these are the absence of a consumer credit or installment purchase system, the paucity of housing and the inadequacy of mortgage financing, and the gradual disintegration of extended families into nuclear families with the attendant loss of a support network for individuals entering retirement.

1.35 The recent behavior of the household savings rate is very encouraging. There has been a steady increase from 6.0% in 1980 to 7.1% in 1983, followed by a large jump to 9.3% in 1984. While this is primarily due to the growth of household incomes since 1980, increases in the real rate of interest have probably been of considerable importance also. Measured from negative levels in 1980, the real rate of interest available to depositors has risen by almost ten percentage points by 1985. This amounts to a substantial change in the opportunity cost of current consumption. Econometric analysis suggests that such a change in real interest rates can produce a sizable increase in the savings rate, by as much as 1.2% of GNP according to one study.[11/]

1.36 Enterprise saving has exhibited remarkable constancy (at around 10%) since 1970. Year-to-year variations appear to be related to income growth which determines profit levels, and possibly also to changes in tax and depreciation incentives. The decline in enterprise savings in 1981 and 1982, for example, appears related to the economic slowdown experienced since 1980. What is worrying, however, is the lack of response of enterprise savings to the rise in the GNP since 1982. This is partly due to the fact that profit levels have risen at a slower rate than GNP.

10/ See, for example, Yusuf and Peters (1984).

11/ This is based on estimates reported by Yusuf and Peters (1984). While their estimates apply to the overall savings rate, the interest rate effect is probably most relevant for the household savings component.

Table 1.6: SAVINGS RATE DEVELOPMENTS
(%)

	1980	1981	1982	1983	1984	1985
National	20.8	20.5	20.9	25.3	27.9	28.4
Private	15.4	14.9	14.8	18.1	20.9	21.4
(Household)	6.6	6.9	6.8	7.6	9.9	21.4
(Enterprises)	8.8	8.2	8.0	10.5	10.0	21.4
Government	5.4	5.6	6.1	7.2	7.1	7.0

/a New SNA methodology.

Source: BOK, Economic Statistics Yearbook.

1.37 The turnaround in business prospects occasioned by the recent sharp drop in the price of oil and the appreciation of the yen may stimulate a higher rate of corporate savings in 1986. In the short run, retained earnings will surely rise as sales increase and costs decrease. In the longer run, wage and salary adjustments and increased competition for sales will work to bring the corporate savings rate back to levels consistent with historical patterns of investment financing in Korea. A permanently higher rate of corporate savings is probably possible only if the pattern of investment financing is fundamentally altered to encourage less reliance on debt and more reliance on retained earnings and equity.

1.38 Government savings have risen in recent years from 5.8% in 1980 to 7.9% in 1984. One would expect the amount of such savings to be determined principally by the level of GNP and the inflation rate. The level of GNP ordinarily determines the level of revenue from taxes and, given constant expenditures, higher revenues would translate into higher government savings. The inflation rate affects government savings in two ways, through bracket creep and through the inflation tax which reduces government debt and transfers resources to the Government. Both of these factors had been operative in the 1970s and led to an increase in the government savings rate from 3.5% in 1972 to almost 7% in 1979. Since 1981, however, inflation has been brought down and it has been increases in the level of income, on the one hand, and tight controls on fiscal expenditure, on the other that have boosted the government savings rate.

1.39 Will government savings continue to grow? There are several reasons to be skeptical of the possibility of further growth in this category. First, Government is committed to a policy of low inflation and thus little can be expected in terms of revenue enhancement from bracket creep or from the inflation tax. Second, there are, at present, no plans to raise tax rates. If anything tax rates have been lowered in recent years to stimulate business.

To the extent also that the pattern of government expenditures envisaged for the Sixth Plan period involves greater emphasis on social development (education, housing, medical care, and social security arrangements), which typically does not produce revenue, the scope for government savings is even lower.

1.40 Higher rates of government savings would be possible if there were to be a higher rate of GNP growth than currently targeted and if significant cost savings could be achieved in government consumption. Both of these possibilities have been enhanced by recent favorable external developments. The decline in the price of oil should enable the presently targeted levels of government expenditure to be achieved at lower cost. Furthermore, higher GNP growth is also possible. Therefore, provided expenditure targets are not revised upwards significantly, a higher rate of government savings appears possible in the foreseeable future.

Assessment

1.41 The main considerations in assessing the likelihood of further growth in the savings rate are that income growth prospects are bright and an appropriate monetary/financial policy is in place. Government is committed to allowing deposit interest rates to move eventually towards the freely determined cost of capital in the economy. This should not only have a direct effect on savings, but a possibly more powerful indirect effect, i.e., by improving investment efficiency, the process of financial liberalization provides greater stimulus to growth of income and this leads to further savings.[12/] It has already been noted that investment efficiency has been rising in recent years as Government has withdrawn progressively from fixing the price and allocation of credit and from project selection. Continuation of this trend should help boost savings. However, an environment of income growth and financial liberalization is the minimum requirement for continued growth of savings. Given such an environment the Sixth Plan target of a 33% savings rate by 1991 may be considered optimistic but not out of reach.[13/] Absent such an environment, however, the target may well be out of reach. The challenge for policymakers is to ensure the continuation of such a favorable environment.

E. External Financing: Market Access and Risk

1.42 The single most important reason for worrying about an uninterrupted supply of foreign loans is that Korea has built up a large external debt on its way to a $2,000 per capita income and therefore faces a large repayment bill. Even if it were to keep its current account in balance for the

---

12/ This view is associated with the work of McKinnon and Shaw and has received empirical confirmation in the case of Korea in Fry (1985).

13/ The schedule of savings rate targets for the Sixth Plan period implies marginal savings rates ranging from 37% to 40%. These are admittedly high but should be assessed in the light of the fact that Korea actually achieved an average marginal savings rates of 36% over the 1981-85 period.

remainder of the decade, Korea would face an annual gross external financing requirement rising from $7 billion in 1986 to $11 billion in 1990. These amounts would be necessary to finance amortizations of the large debts piled up in the early 1980s, rising exports on a deferred payment basis, and increases in reserves to match higher trade flows in the future. Of this about $7-8 billion will have to be raised from private sources, principally banks.

1.43 Lending by foreign commercial banks grew at a tremendous pace during the 1970s and early 1980s as Korea sought funds to cope with the oil price shocks and banks sought eagerly to recycle their petrodollar deposits. Korea's debt grew from $2.3 billion to $45 billion during 1970-85 and from $27 billion to $37 billion in just the two years 1981 and 1982. Commercial banks have financed over three quarters of recent borrowing. The history of Korea's debt accumulation indicates that two sets of factors influence the supply of foreign loans, factors external to the country which have to do with the global strategies and circumstances of the commercial banks and factors internal to the country which affect its creditworthiness in the eyes of its financiers. An overview of the likely development of such external and internal factors over the rest of the decade suggests that the two sets will influence the supply of loans to Korea in opposite directions. External factors are shaping up in such a way that a tightening of loan supplies is indicated. Internal factors, however, and principally the excellent management of the economy that has been evident in recent years, indicate that Korea will maintain its creditworthiness and therefore should attract adequate funds. These factors are discussed in turn in the following sections.

Determinants of Loan Supply

1.44 Many of the US banks which have provided the bulk of bank credit to Korea in the past are now faced with binding country exposure limits. Portfolio considerations require them to reduce the growth of Korean assets on their books. Such a trend has been evident since 1981. Net lending to Korea by US banks declined from $2.3 billion in 1981 to $400 million in 1983. In 1984 and 1985 the flow was actually reversed and average net lending amounted to -$1.25 billion. The consensus remains that the prospects for significant positive net lending by US banks are slight. While the slack was adequately taken up by Japanese and European banks in 1984 and 1985, there is concern that binding exposure limits are rapidly being reached for many of these also. Greater diversification among providers of syndicated loans does not seem to be a readily available option, certainly not at the terms to which Korea has become accustomed.

1.45 There has been a change in the regulatory atmosphere for US banks which, by requiring, inter alia, increased capital to asset ratios, discourages the taking of risks that were considered reasonable in the seventies. The Latin American debt experience and recent troubles with energy and real estate loans have chastened both US banks and their regulators. They are now inclined to err on the side of excess caution; ergo, there is now a greater concern with asset quality than before. There may also be a zero-sum game aspect to future international finance arising from the present debt crisis: involuntary lending to Latin America, combined with stricter regulations and

Table 1.7: INTERNATIONAL BANK NET LENDING TO KOREA /a

	1981	1982	1983	1984	1985
Net Lending ($ bln)					
US banks	2.3	1.5	0.4	-1.5	-1.0
Other banks	1.1	2.0	1.8	3.5	n.a
Total	3.4	3.5	2.2	2.0	
Growth of Claims (%)/b					
US banks	32	16	4	-13	-9
Other banks	14	24	17	30	n.a.
Total	23	20	10	9	

/a Changes in cross-border claims adjusted for exchange rate changes.
/b Net lending as a proportion of the stock of claims.

Source: Country Exposure Lending Survey, Federal Financial Industries Examination Council, Federal Reserve Board.

conservative exposure limits for developing countries as a whole may mean a reduction in credit available even to the better risks among developing countries, such as Korea. Moreover, financial deregulation has, in recent years, also increased opportunities for profitable lending in the OECD countries and especially in the US. One reason why US banks may no longer want to lend as much abroad as before is because they see relatively better opportunities at home. So, despite its good track record, Korean paper may not be as attractive an asset.[14/]

1.46 The fiscal stance of the US Government is another cause of international financial imbalance. Large US deficits are being financed with capital from OECD countries (and also perhaps with reverse capital flows from some LDCs). This has both raised interest rates and decreased the supply of loanable funds to such countries as Korea. Until the US fiscal deficit is reduced, its financing will continue to have a crowding out effect on a world scale. The situation was different in the 1970s when the US was a capital exporter but now that it has become a capital importer and is likely to continue to remain one for some time to come, prospects for the availability of capital for other countries must be reassessed. Together with financial deregulation this has caused a substantial reallocation of credit away from LDCs and towards the bonds and equities of the US.

14/ Korea's good track record and continuing creditworthiness is reflected in the low spreads at which it can borrow from international loan and bond markets, which paradoxically may also lower its attractiveness as a new asset to some banks.

1.47 The preceding section has indicated that external developments are likely to push for a tightening of the supply of loans to LDCs, Korea included. Nevertheless, bankers are united in their view that Korea's prospects remain the best under the circumstances because of its excellent economic management and that Korea would be the first among the LDCs to benefit from improved global liquidity. Furthermore, Korea can expand its access to loans by undertaking measures to attract a wider range of lenders.

Sources of Risk in Debt Structure

1.48 An uninterrupted supply of foreign loans cannot be taken for granted because it is partly determined by factors external to Korea. In addition to this access risk there are other sources of vulnerability in Korea's debt situation. It is convenient to describe these additional risks separately under the headings (a) maturity risk, (b) interest rate risk; and (c) concentration risk. The risks are discussed below and quantified to some extent in Table 1.8.

1.49 The maturity risk arise from the fact that a significant fraction (26%) of Korea's debt is short term. Most of this short-term debt is trade-related but a non-negligible portion is also used to finance long-term investments. If Korea is to maintain its growth it must rely on trade and trade finance. A sharp cutback in trade finance for whatever reason can affect Korea's growth path substantially. To the extent that some short-term debt goes to finance longer-term investment projects the damage to growth could be even more severe. Conventional rules of thumb suggest that short-term debt should not exceed the value of three months of imports. In Korea's case, short-term debt has exceeded this level for the past five years but the gap has been declining. Korea has taken steps to improve the maturity structure of its debt in recent years and has met with success. Short-term debt has been reduced from 34% of external debt in 1980 to 26% in 1984.

1.50 It is in Korea's interest to continue making efforts to improve the maturity structure of its debt and, in particular, to promote use of long-term funds to finance long-term projects. One way to do this is to encourage greater use of official sources of international finance which typically offer longer maturities and grace periods than do commercial banks. While this may be a costlier option in some instances the extra cost should be balanced against the risk-reduction achieved.

1.51 Interest rate risk arises from the fact that a large proportion of Korea's debt is in the form of floating rate loans (and bonds). This increases the repayment obligation at times when interest rates are rising and new loans are becoming more scarce. While Korea benefits from variable rate loans at times when interest rates are dropping (as at present) there may be advantages in reducing the proportion of debt denominated in variable rates. This can be done either by negotiating fixed rate loans (or bonds) directly or by swapping a floating rate obligation into a fixed rate one. The costs and benefits of changing the proportion of fixed and floating rate obligations will depend on the terms available in the market. It is safe to assert, however, that at present too large a proportion of Korea's debt is in flexible-rate form. While it may seem risky to engage in swaps, the current vulnerability of the debt portfolio to interest rate shocks is in-and-of-itself a substantial risk.

Table 1.8: BORROWING: RISK INDICATORS

	1981	1982	1983	1984	1985
Maturity Risk					
Short-term debt/total debt (%)	31.5	33.5	30.1	26.8	23.0
International reserves in import months	3.6	3.9	3.6	3.6	n.a.
Short-term debt in import months	4.7	6.1	5.1	4.5	4.1
International reserves/short-term debt (%)	70.4	57.9	61.8	72.0	n.a.
Interest Rate Risk					
Interest payments/exports (%)	12.8	12.8	10.5	11.2	11.2
Variable rate loans/total loans (%)	53.5	56.6	53.9	53.5	n.a.
Concentration Risk					
Debt/total capital inflow (%)	99.0	98.4	98.1	94.6	93.6
Debt/GDP (%)	47.2	51.3	52.5	51.1	54.2
Debt service ratio (%)	21.2	22.6	20.9	22.6	21.8
Official source debt/total debt (%)	40.4	42.2	41.2	39.3	n.a.
Guaranteed debt/total debt (%)	85.3	85.3	82.2	82.3	n.a.

Source: World Bank Debtor Reporting Survey.

1.52 Perhaps the most important risk in Korea's debt arises from the fact that it is extremely concentrated in the form of conventional bank loans. At present over 90% of foreign capital inflow into Korea is in the form of bank loans and these come primarily from US and Japanese banks.[15/] Thus, at the moment, Korea has a highly unbalanced portfolio of external liabilities. This portfolio has several disadvantages. First, it does not provide Korea with a hedge. When world economic conditions deteriorate and Korea's situation does likewise it scope for remedial macroeconomic action is reduced by the necessity of making large payments at fixed intervals regardless of its current condition. In fact, such large payments probably amplify the domestic distress associated with the economic downturn. An obligation whose liability adjusted to Korea's economic health would be of considerable advantage.

15/ In recent years Korea has raised increasing amounts of external finance by floating bonds. Whereas bond-type instruments accounted for only 6% of gross annual external financing in 1982, their share rose to about 24% in 1984. However, Korean bonds have several characteristics that make them very similar to conventional medium-term loans (e.g., many carry four-year "put" options). Moreover, they have been purchased largely by US and Japanese banks and hence they have not provided much diversification by source.

1.53 Second, bank loans are a general obligation credit and do not help in spreading either the risk or the responsibility of the project or projects that they finance. The risk of project failure is borne not by the financier but by the borrower. Also, there is no particular incentive for the financier to ensure that the project is suitably designed and executed to maximize the probability of success. The unwritten rules of the game governing lending to LDCs in fact have the entire loan placed at sovereign risk regardless of whether an actual guarantee is extended by the sovereign. An instrument which shifted risk in accordance with comparative risk-bearing ability among borrowers and lenders would be of considerable advantage. To the extent also that such instruments created incentives for responsibility-sharing, they would be useful in ensuring that good projects are undertaken. Equity instruments accomplish these tasks far better than debt instruments.

Role of Non-debt Instruments

1.54 Equity financing can occur in two ways, through direct foreign investment (DFI) and through portfolio foreign investment (PFI) or the purchase of Korean stocks by foreigners. These have not been major sources of financing in the past. DFI has, for example, never amounted to more than 10% of annual capital inflow and has usually been less than 5%. PFI was virtually nonexistent until the early 1980s when Korea first allowed purchases of Korean stocks through mutual funds. Reforms have taken place; however, there are still numerous impediments and disincentives to foreign investment.

1.55 Now may be an especially opportune time to engage in external liability diversification through the promotion of direct and portfolio foreign investment. The drop in the price of oil and interest rates has stimulated stock markets the world over. Share prices are rising in Korea and companies have a chance of issuing at very favorable rates. Foreign investor interest in Korean securities is high, as evident, for example, from the rapid increase in the market price of shares of the various stock funds available to them. Several Korean companies have become well known abroad on account of their exports. In particular, the conglomerates engaged in the manufacture of automobiles and electronics are establishing a brand-name presence in foreign markets and should, therefore, attract the interest of a wide range of investors, including non-institutional ones. Moreover, large OECD institutional investors are highly liquid at present as a result partly of several years of cost cutting and recovery in sales from the recession of the early 1980s. Intense merger activity in 1984 and 1985 were indications of such liquidity. The decline in oil prices should boost corporate liquidity even more as costs fall further. For all these reasons, the near future may be an excellent time to draw foreign investment into Korea.[16]

16/ Among the steps that Government might consider by way of accelerating the pace of internationalization of the equity market are: (a) substantial expansion of existing investment trusts for foreigners; (b) establishment of additional funds, similar to the Korea Fund, on other foreign stock exchanges such as the Tokyo exchange and various European stock exchanges; and (c) increasing the number and size of convertible debenture offerings permitted to Korean corporations, actions entirely in keeping with the thrust of announced Government policy with respect to capital market development.

1.56 The encouragement of a greater role for direct foreign investment in Korea should help increase competition and efficiency since Korea's emerging industries (e.g., precision machinery, automobiles and electronics, among others) require not just off-the-shelf foreign technology, but also foreign management expertise. In the past, Korea exports such as textiles and footwear were characterized by markets in which production efficiency was most important and the technologies, for the most part could be delivered on a turnkey basis and operated without much need for design and capacity innovation. Gradually, Korea has moved out into more complex technical processes and while a remarkable capacity to master these also has been demonstrated there remain areas of weakness, partly because of a shortage of the requisite skills and partly because of weak corporate financial structures. Korea's ability to advance in these areas may be significantly improved by the encouragement of direct foreign investment.

1.57 When oil prices rose sharply through the 1970s, much of OPEC's excess wealth was recycled through large international banks. Now that oil prices are falling, a different sort of recycling may be in the making, involving large OECD corporations rather than banks as the primary agents. The direct benefits of the oil price decline will accrue principally to the OECD nonfinancial corporate sector, rather than to OPEC governments as was the case when oil prices rose. These entities are likely to have different portfolio allocation strategies. The principal gainers will be Japanese and European companies who have limited domestic investment opportunities. Japan and Germany, for example, are surplus countries which have been actively seeking to invest their savings abroad, and Korea would provide a promising opportunity. While investment in Europe continues to be hamstrung by the existence of entrenched and powerful labor unions, such is not the case with Korea. Should foreign investment controls be relaxed, a boom in investment could follow.

1.58 Equity investment should also be considered for its potential role in helping Korea's exports. Direct foreign investment could bring with it marketing and design capability that is needed if Korea is to expand its share of exports in such items as autos, specialty machines, high quality textiles and electronics. Foreign investment could also help in relieving the pressure of protectionism. The larger the stake that OECD corporations have in Korea's industrial growth the greater will be their aversion to protectionism in their home markets. Korea would do well to recruit major corporate allies in its campaign against protectionism, thus direct foreign investment may have a role to play in Korea's trade diplomacy as well.

F. The Medium-Term Outlook

1.59 Korea has established ambitious targets in exports, savings, external debt reduction, and income growth for the remainder of the 1980s. These targets face uncertainties, only some of which are amenable to domestic policy action. The principal risks are external. Difficulties would ensue if a world recession were to be encountered, if protectionism against Korean exports were intensified, or if the supply of foreign loans were drastically curtailed. All three developments would jeopardize export and savings targets and produce lower growth as well as continued vulnerability to changes in international financial markets. The occurrence of one or more of these downside risks would place Government on the horns of a dilemma as it would have to sacrifice one or more of its targets.

1.60 The most important tradeoff would arise between income growth and external debt reduction. For example, if the supply of foreign loans were curtailed, Korea would be faced with a choice between either maintaining investment through the inflow of other forms of capital (such as equity investments) or slowing down its rate of growth to a level consistent with the reduced level of external financing. Similarly, a cyclical downturn in the world economy would affect Korea's export growth and thereby threaten its growth objective. Korea would then be faced with a choice of either maintaining income growth through domestic expansion, but at the cost of higher inflation and higher external debt, or of accepting the growth reduction. The latter choice might create further problems for the accelerated debt repayment envisioned in the Plan. Korea's successful pursuit in recent years of the multiple goals of higher savings, higher exports and higher growth together with price stability should not blind it to the inherent difficulties of such a task.[17/] If the external environment is not as accommodating in the future as it has been in the recent past (1985 excepted), these difficulties will become readily apparent in the form of more difficult tradeoffs.

Linkages and Lessons

1.61 A strong interrelationship exists between income growth, external debt and domestic savings. Hence, policies to change any one of these variables cannot be considered in isolation. For example, since it is income growth, and particularly export-led growth, that generates the resources with which to reduce outstanding external debt, it is not necessarily prudent to choose a low debt acquisition route if that route implies lower growth. If debt reduction were to result in a reduction in income growth this could depress domestic savings and possibly worsen the investment-savings gap. Such a strategy would not be successful in reducing debt reliance and could be counterproductive, particularly if the proximate cause of debt increase is transitory. If productive investments exist that have a rate of return higher than the marginal cost of external debt, it makes economic sense to acquire the incremental debt. In the ultimate analysis it is not the level of debt that is critical but the efficiency with which it is used. Furthermore, a direct reduction of debt is not necessarily the best way to reduce financial vulnerability. In some cases this can be achieved by diversifying among sources of foreign capital, e.g., away from debt towards equity instruments. This provides a perspective within which to discuss possible macroeconomic responses to shocks that may affect Korea's export or savings performance in the future. The discussion draws on some recent features of Korean macroeconomic policy, elaborated on in Chapter 2, Volume II.

17/ In a reformulation of Tinbergen's famous axiom, the Koreans are proud of having "caught three rabbits with two hands," namely, simultaneously achieving growth, inflation, and payments objectives.

1.62 In quantitative terms the most significant shock to exports is likely to occur if foreign income growth were to slow down. If such a shock occurs, what should the appropriate policy response be? The policy experience of 1985 provides some guidance. Stagnating OECD incomes depressed demand for Korean exports in late 1984 and raised the possibility of a sharp deterioration in export growth. Korea reacted by allowing growth to fall below target but not to collapse. The policies employed were principally a sharp depreciation of the won and somewhat easier domestic monetary policy. While both policies helped to keep economic growth from falling off too sharply, the depreciation helped to keep the balance of payments performance on target. Higher inflation was to be expected as a consequence, but the fall in oil prices probably offset imported inflation.

1.63 All in all, the 1985 outcome was very successful for Korea as it managed to both keep income growth from collapsing without reigniting inflation and squeeze a little more export growth and an increased share out of a slow market. Future foreign shocks should probably be handled the same way, i.e., by a domestic expansion which bolsters growth but is not completely offsetting. Given that oil and commodity prices are forecast to stay soft over the medium term, the downside inflationary risk of such a strategy remains low, although it must be noted that an overly vigorous domestic expansion which raises inflation can have a depressing effect on domestic savings and can worsen the current account balance.

1.64 If the source of a deteriorating balance of payments situation is decreasing competitiveness caused by high domestic wages, for example, the appropriate policy response would be the one followed after 1979-81--a reduction in aggregate demand through contractionary macroeconomic policy combined with a devaluation. Both policies would, in the Korean context, reduce import demand while the latter would promote exports quickly and perhaps offset the adverse effects of the former on employment. The challenge would be to limit the rise in real wages without slowing down output growth drastically. If decreasing competitiveness is industry-specific, macroeconomic policies will not supply an appropriate long-run remedy. More appropriate would be policies designed to ease the exit of firms and workers out of the declining industry and into a growing one at minimum social cost. This probably applies to the case of such industries as textiles, and overseas construction at present, and possibly shipbuilding in the future if no global turnaround materializes. Macroeconomic policies are ill-suited to deal with the problems of structural decline.

1.65 The point of the foregoing is that a decline in exports should not automatically trigger an expansionary macroeconomic policy. The proximate causes of the decline must be clearly identified and policy framed accordingly. In addition it is important to re-examine the importance of the export or BOP target in different circumstances. If it is clear from the circumstances that the deterioration is cyclical or transitory, a hands-off policy response would be appropriate and the deterioration should be accommodated by an increase in debt. Such an option is of course difficult if the country does not have access to financial markets to cover its deficit, but Korea faces no such difficulty. The implication, however, is that self-imposed constraints on debt acquisition may have to be breached and vulnerability to international financial shocks increased.

Overall Assessment

1.66 The Sixth Five-Year Plan is ambitious, yet achievable. It relies in good measure on a more favorable external environment in the late 1980s, of which 1986 has begun auspiciously. There may well be setbacks, and the ultimate achievement of key target variables will depend, as in the past, on swift, consistent, and pragmatic policy responses. There are new risks on the horizon, as well as new opportunities. The risks are embodied in the new products and markets on which Korea is placing great emphasis and on the investments behind that modern export strategy. Risks are also borne by the markets supplying the factors of production in a world of rapidly changing comparative advantage. More rapid employment shifts reflecting changing skill mix requirements are to be expected, and either must be accommodated with a minimum of friction (with attendant social issues to be faced by Government) or resisted, clearly a less desirable course. Greater risks will also be borne by capital markets, which points to the need for greater resiliency on their part. In large measure, therefore, the achievement of the Sixth's Plan rests with the success of Korea's desired industrial transformation.[18/]

1.67 As already indicated, the achievement of Fifth Plan targets were as materially facilitated by farsighted supply-side interventions as by prudent demand management policies. Actions to improve the efficiency of markets and their competitiveness paid handsome dividends. For the Sixth Plan's success, these supply-side actions must be augmented. Market imperfections which were tolerable at earlier stages of development, particularly when market offsets were more easily handled via direct public interventions, are now likely to be less effective and, moreover, less desirable. Therefore, as is discussed in the subsequent chapters, economic reforms in the areas of trade, industrial finance, and industrial policy will be critical in (a) sharpening the economy's ability to respond to changing market opportunities, and (b) reducing the drag on the economy caused by lackluster industries and ill-functioning markets. If these reforms are successfully implemented, Sixth Plan targets stand a very good chance of success.

G. The Current Economic Outlook

Implications of Recent Developments

1.68 The external environment presently features favorable developments in both oil prices and interest rates. Should these continue through the medium run, they will assist Korea in expanding exports and achieving a surplus on the current account, thereby raising the growth rate while simultaneously reducing its reliance on external debt. The current situation is reminiscent in some respects of the situation that existed in late 1983 when the Fifth Plan was revised. Now as then the domestic rate of inflation and the current account deficit are under control and GNP growth is on track. Externally, oil prices and interest rates are declining. There are, of course,

---

18/ See Chapter 1, Volume II on Korea's changing industrial structure for an identification of future issues.

several respects in which the current situation is different. The recent decline in oil prices and interest rates has been much faster and of greater magnitude than the trend observed during 1984-85. Additionally, the Japanese yen has appreciated significantly against the won since December 1985. The simultaneous occurrence of those developments promises to deliver significant short-run benefits to Korea.

1.69 The medium-term implications of these events appear to be quite favorable at this point as well. Oil prices are forecast to stay soft over the next five years. The yen is forecast to remain strong relative to the dollar (and the won) on account of Japan's current and expected trade surpluses and because of international pressure on Japan to refrain from taking domestic policy measures that would offset the momentum towards yen appreciation. The future evolution of interest rates is harder to predict. It will be strongly influenced by the behavior of the US budget deficit and by the nature of the eventual resolution of internal debt problems in the farm and energy sectors of the US and the international debt problems of the Latin American countries.

1.70 These recent events have significant implications for macroeconomic management in Korea. Broadly speaking, the task of management will become easier since various constraints on the economy will be relaxed. Both monetary and fiscal policy can be more expansionary than has been the case in recent years without fear of significant adverse effects on either inflation or the current account. The dominant influences on domestic inflation will continue to be the import price index and the rate of growth of domestic wages. Softness in oil and commodity prices will continue to exert downward pressure on the former, and, although the appreciation of the yen will exert some upward pressure (since Japanese imports are important to Korea's export growth), the net effect is expected to be deflationary.

1.71 Domestic wages may begin to grow faster as investment accelerates and as workers attempt to compensate for the belt-tightening of recent years. However, the domestic labor market is currently slack and may get slacker still with a sharper decline in overseas construction activity. On balance, domestic wage growth may not be a source of great pressure on inflation in the coming years. Overall, therefore, there exists a significant possibility of negligible inflation in Korea in the 1986-87 period. The prospect for external balance is also brighter, in part as a consequence of Korea's terms of trade, which are likely to improve as the effects of declining oil and commodity prices should outweigh those of the appreciated yen. A significant improvement in the terms of trade makes it less important that an aggressive, competitiveness-oriented nominal exchange rate policy be followed.

1.72 While recent external developments have reduced the downside risks of expansionary macroeconomic policy, they may have, at the same time, reduced the necessity of such a policy stance. Export growth, stimulated by the growth of external demand, may provide sufficient impetus to keep the Korean economy growing at the 7.5% that is thought to be consistent with (present) Sixth Plan targets for monetary and fiscal expenditure growth. It is only if the anticipated export growth does not materialize or if a higher growth rate of GNP is deemed necessary (perhaps to meet a revised employment target) that the need for more expansionary macroeconomic policy should arise. In such a circumstance, the greater latitude now available to policymakers will be useful.

A Window of Opportunity

1.73 Since 1979, a process of structural adjustment has been underway. This process featured, *inter alia*, initiatives to change the nature of government intervention in industrial matters, to liberalize imports, to reduce government control over credit allocation and pricing, to reduce the reliance of Korean corporations on debt, to promote the domestic capital market and to internationalize the Korean financial and industrial sectors by allowing greater foreign participation. While some progress has been achieved in each of these areas, the pace of reform has, in some instances, been slower than desired. The content of trade liberalization was, for example, influenced by concerns about the current account deficit. The pace of financial liberalization, especially in the matter of interest rate decontrol, was affected by concerns for the health of highly-leveraged corporations facing poor profit prospects. The speed of capital market internationalization was constrained by the undervaluation of equity and the apprehension that weak domestic corporations would not be able to negotiate fair bargains with strong foreign corporations. Finally, the process of industrial policy reform has been affected by the unsteady state of certain industries and the constrained finances of the Government.

1.74 The situation has now changed considerably. Many of the reforms initiated in the 1979-81 period are now paying off. Moreover, because the sharp decline in oil prices and interest rates should make 1986 a banner year for Korea, it provides policymakers with an ideal opportunity to both consolidate and augment these gains. The threat of inflation and rising external deficits is considerably reduced. Exporters face improving markets with a highly competitive won. Government finances as well as bank and corporate profits should improve substantially. Trading activity and the average price per share in the domestic capital market have increased significantly in early 1986. Thus, the current conjuncture of favorable developments presents a rare opportunity to accelerate the process of reform and liberalization and provide the economy with the additional flexibility and resilience it needs if Korea is to achieve its medium and longer-term objectives.

CHAPTER 2: THE LEGACY OF KOREA'S INDUSTRIAL POLICY

A. The Issues of Industrial Policy

2.01 Korean industrial policy is notable for the prominent role of Government in the economy, the boldness of policy changes, and, not least, extraordinary results. A review of this remarkable legacy helps clarify present policies--such as biases in credit access or restrictions on imports--in terms of the historical context in which they evolved. It also helps to explain the evolution of prominent features of the Korean economy that affect present policymaking, including its key strengths, such as adaptibility to changes in world markets, as well as its problems, such as relatively immature capital markets.

2.02 There is little doubt about either the main directions or chief accomplishments of Korea's development strategy. While Korea's export "takeoff" started from an unusually low base, and was supported by a general expansion of world trade, it would not have been possible without decisive and innovative policies. These included a rationalized exchange rate regime, strong export incentives, selective import liberalization, directed credit, and a host of finely-tuned, export-promoting instruments.

2.03 Aggressive export orientation emerged as the central theme of Korean policy in the late 1950s and has remained so since. Export growth climbed into the 20-40% range in the early 1960s and remained high with few interruptions until the late 1970s. Korea was the world's 101st largest exporter in 1960 and is the world's 13th largest today; its average rate of export growth between 1960 and 1983 was twice that of Japan. Growth was especially rapid from about 1964 to the mid-1970s--a period that witnessed significant steps in export promotion and some trade liberalization.[1/]

2.04 As its exports expanded, Korea moved from a largely insular economy to an unusually outward-oriented one. Importantly, outward orientation was not limited to increased exports; the development program included the opening of the economy toward imports and capital inflows as well. Foreign borrowing helped to boost capital formation significantly above national saving. That is not to diminish the critical contribution of Korean savings, which rose from a small proportion of gross domestic capital formation in 1960 to 50% in 1965, 65% in 1970 and 92% in 1973 (see Figure 2.1). A unique feature of Korean industrial development in the early years was the efficiency with which the economy utilized foreign markets, products, and resources to generate domestic growth.

1/ The scope of this export growth is illustrated by comparing Korean export patterns with typical values calculated from the cross-country analysis of Chenery and Syrquin (1975). Korea was well below international norms in the 1950s. It then passed the norms in the mid-1960s, and by 1980 Korean exports and imports were several times the levels predicted by cross-country analysis. See Chapter 1, Volume II for further details.

Figure 2.1

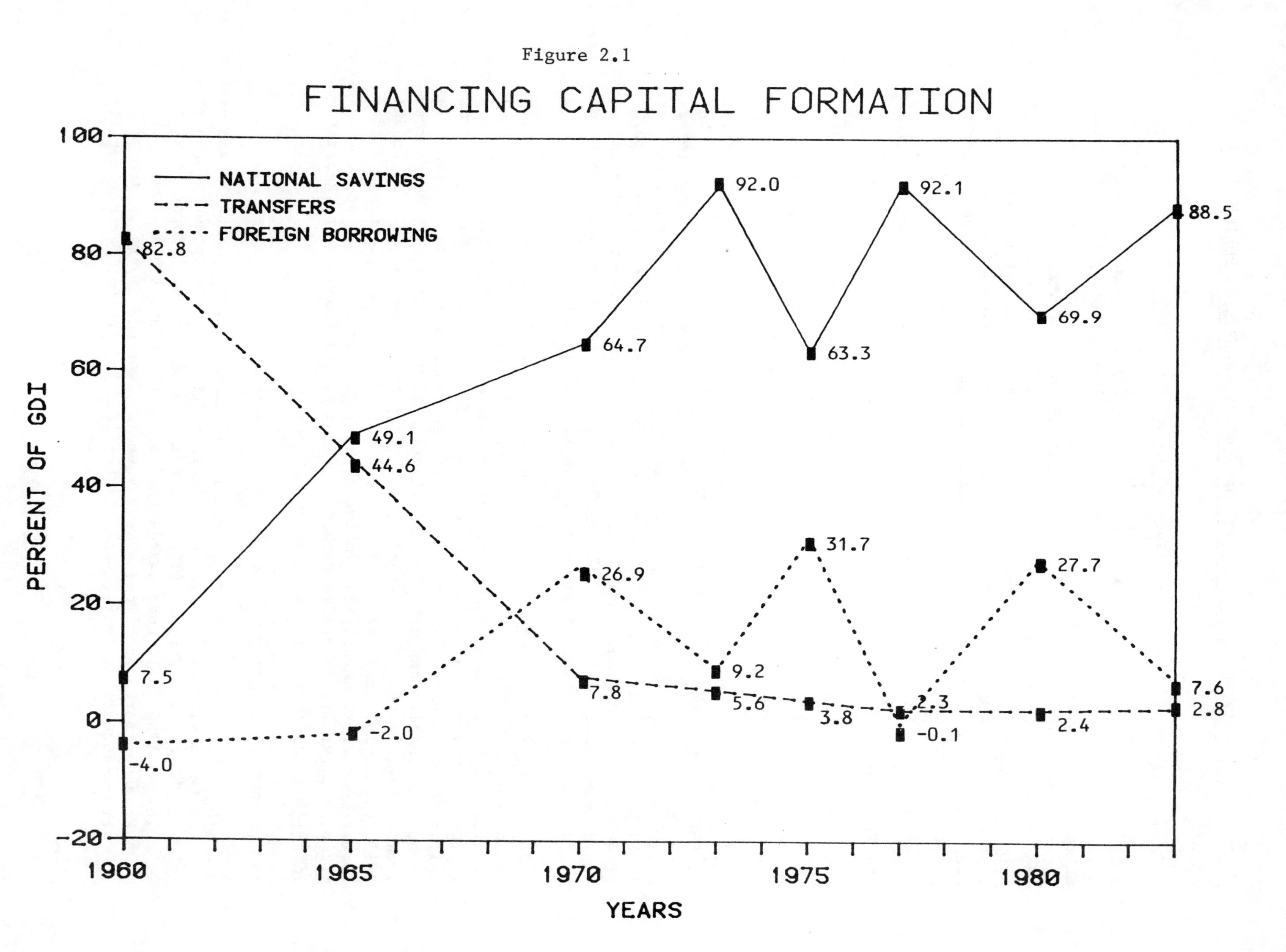
FINANCING CAPITAL FORMATION
NATIONAL SAVINGS
TRANSFERS
FOREIGN BORROWING
PERCENT OF GDI
YEARS
100
80
60
40
20
0
-20
1960
1965
1970
1975
1980
82.8
7.5
-4.0
49.1
44.6
-2.0
64.7
26.9
7.8
92.0
9.2
5.6
63.3
31.7
3.8
92.1
2.3
-0.1
69.9
27.7
2.4
88.5
7.6
2.8

2.05 Contemporary development theory argues that an effective growth strategy requires a direct link between profitability in world markets and domestic incentives. In important ways, Korean policy established such a link at a very early stage. But Korean strategy did not wholly conform to theoretical prescriptions. Government intervention exceeded the level generally held to be appropriate. And while some crucial prices were "gotten right," others remained distorted. Only in its most recent phase has Korean industrial policy begun to emphasize generalized liberalization as a central objective.

2.06 Korea's industrial policy history can be roughly divided into three phases: (i) the "takeoff" phase between 1961 and 1973; (ii) the heavy industry promotion phase between 1973 and 1979; and (iii) the liberalization phase since 1979. The takeoff phase witnessed wide-ranging interventions in export promotion, industry finance, and protection. As shown below, this complex policy regime succeeded in part because intervention was guided by clear criteria of economic efficiency.

2.07 The Heavy and Chemical Industry (HCI) drive, which roughly began in 1973 and lasted until 1979, involved more typical sector-oriented, import-substituting kinds of initiatives. Designed in part to accelerate changes in Korean comparative advantage, the HCI drive provided extensive support to large-scale, capital-intensive industries. Some of the production and export objectives of the plan were eventually realized, but its dominant effects--underutilized capacity, "crowding out" of traditional export industries, and a sharp decline in the incremental capital output ratio--were unfortunately reminiscent of experiences elsewhere.

2.08 Since 1979 Korea has pursued a slow, but deliberate policy of liberalization. In contrast to the liberalization experiences in South America, there is little urgency or drama to this effort. The Government is withdrawing slowly, although unevenly, in the policy areas of domestic finance, import barriers, and direct export promotion. So far, the experiment is yielding positive results. If current policies remain on track and work, Korea's economic policies will eventually become as much a model for successful liberalization as its past policies now are for successful export promotion.

2.09 The key elements of Korean industrial policy included exchange rate and trade policy (since these fundamentally affect incentives across different branches of the economy), financial policies that governed the allocation of capital, and direct government actions that affected industrial structure and infrastructure, and especially technology. These elements are examined below in the context of the roles they played in different periods of Korean development. The concluding sections of the chapter assess the reasons for long-term trends in policies, and their legacy for contemporary decision-making.

B. Managing the Takeoff: 1961-1973

Policy Framework

2.10 The 1961-1973 period represented a unique combination of aggressive orientation and classic import protection. Korean policymakers had clear control over trade, exchange, and financial policy, as well as aspects of industrial decision-making. In contrast to other countries, however, they used these instruments in an integrated fashion to pursue the primary objective of export growth. Industrial policy during this period involved neither functional interventions (addressing specific types of market failure) nor selective interventions (influencing the industry-specific composition of economic activity), but rather a comprehensive incentive system designed to channel resources into export-oriented activities. At the same time, the period witnessed gradual changes in policy, which slowly replaced the more bureaucratic market-steering interventions with market-enhancing incentive schemes.

2.11 The development program of the early 1960s was based on the philosophy that accelerated investment, growth, and economic independence required discipline and sacrifice. One application of this principle was the continuation of import controls on consumer and other goods not deemed to be necessities. But the protection of domestic markets also tended to increase profits in the domestic markets and direct domestic resources into import substitution rather than export activities. Whether consciously or not, the government countered this tendency by developing a complex and innovative array of export promotion policies. As a consequence of aggressive export promotion, export industries faced less variation in incentives than occurred in real exchange rates.[2/]

Trade and Exchange Rate Policy

2.12 The key precondition to outward oriented growth was the reform of the exchange rate regime. In the late 1950s, on the eve of outward-looking development, the Korean exchange rate was clearly overvalued. The elimination of this distortion became the logical priority of early reforms, and the major devaluations of 1961 and 1964, along with other supporting policies, provided a strong initial impetus for export growth. Since the real effects of the 1960 devaluation were quickly eroded by domestic inflations, it was particularly important that the 1964 devaluation also inaugurated a sliding-peg system of continued adjustment. Export performance was keenly responsive to exchange rate swings; in the 7 years when the real exchange rate was high in 1965 purchasing power parity terms the average growth rate of exports was 43.5%; in the 7 years when it was low, the average growth rate of exports was 16.0% (see Table 2.1). Importantly, however, the fluctuations in the real exchange rate between 1964 and 1973 were moderate in comparison to those

---

2/ See Frank, Westphal and Kim (1975).

experienced by most developing countries.[3] In other words, by the mid-1960s the Government had established an exchange rate regime that generally conformed to the economy's long-term requirements. This was critical to the success of the outward-oriented strategy.

Table 2.1: EXCHANGE RATES AND EXPORT GROWTH

Real exchange rate (1965 won/$)	Average export growth rate	Years included
<200	16.0	1960, 1963, 1978-82
200-250	30.3	1961-62, 1964, 1967-70, 1976-77, 1983
>250	43.5	1965-66, 1971-75

2.13 The first instruments of export promotion, in force primarily during the 1950s and 1960s, were highly discretionary. Exporters were supported with multiple exchange rates, direct cash payments, permission to retain foreign exchange earnings for private use or resale, and the privilege to borrow in foreign currencies and to import restricted commodities under the so-called export-import link. This system granted exporters access not only to foreign machinery and intermediate inputs for their own use, but also to scarcity rents in heavily protected domestic markets.[4] The export-import link was officially terminated in 1965, but a company continued to have to meet export performance requirements to qualify as an importer until 1975. Even as discretionary incentives were gradually replaced by more automatic instruments, exporters received significant concessions: income taxes on earnings from exports were reduced by 50% (1961), exports and intermediate inputs into exports were exempted from domestic indirect taxes (1961), and exporters were allowed accelerated depreciation (1966). A formal system of wastage allowances permitted exporters to import, on preferential terms, greater amounts of intermediate inputs than required in production (1965). Interest rates on export loans were reduced, and the terms for obtaining working capital loans were improved. Two significant instruments of export

---

3/ It should be noted in this context that Korea's real effective exchange rate has tended towards much less than average variability among middle-income LDCs since then as well. See Edwards and Ng (1985) for a comparison over the 1971-84 period.

4/ The Government began to "mop" up the value of the quota rents in 1961 with administratively determined special tariffs and gradually restricted the granting of "linked" import privileges. Nevertheless, the relatively low level of tariff collections and the discretionary nature of special tariffs suggest that favored exporters continued to benefit from the system until much later. Westphal and Kim (1977), for example, find that cross-subsidization of exports with high profits from imports remained important in 1968, when some trade liberalization had already occurred.

development, still in use today, emerged at this time: tariff exemptions to exporters (1961) and to indirect exporters (1965). These interventions allowed Korean exporters to avoid to a large degree the distortions involved in the protection of domestic markets and thus curtailed the economic losses inherent in the array of import restrictions as well as provided an automatic, off-budget alternative to increasing the subsidization of exporting firms.

2.14 The slowest-emerging component of Korea's outward-looking strategy was the selective liberalization of imports, although Government did recognize that high-cost or low-quality domestic inputs would injure the export effort. Tariffs, and especially import controls, had to be rationalized and contained to encourage the flow of resources into internationally competitive activities. The import system was therefore reorganized, but liberalization played a secondary role to export promotion. Before 1967 most commodities could not be imported into Korea (except for direct or indirect export) unless the industry association representing import-competing firms certified the absence of adequate domestic substitutes.

2.15 While protection could have sharply distorted domestic production incentives, the actual effects were surprisingly benign. Three factors minimized the impact of pervasive controls. First, exporters and indirect exporters largely escaped them.[5/] This was an important effect: the tariff exemptions granted were typically larger than tariffs actually paid. Second, the tariff-equivalent effects of both tariffs and nontariff controls were modest: in 1968, for example, actual tariffs were estimated to be a 9% of import value, and the tariff equivalent of nontariff barriers to be 4%.[6/] Third, at least until the mid 1970s, protection was focussed on final consumer goods and therefore had less effect on productive efficiency than it might have had if targeted on capital or intermediate goods.

2.16 This is not to say that the trade regime was neutral. High levels of protection were afforded to some domestic markets (see Table 2.2).[7/] But analyzed in detail, these variations were also less disruptive than might have been expected. For example, when attention is focused on the difference between incentives for export and domestic sales in the exportables industries, Korea ranks highest (see Table 2.3). This reinforces the view that

---

5/ The exemption of direct and indirect exporters from protection had a marked impact on the composition of exports. The relative profitability of different types of exports was determined by foreign relative prices and comparative advantage. As a result, important export positions emerged in industries that no planner would have targeted, such as wigs. More generally, the early export structure sharply favored labor-intensive goods, with the result that export capacity could be expanded very rapidly despite the scarcity of investment capital.

6/ Westphal and Kim (1982).

7/ According to Westphal and Kim's (1977) data for 1968, import competing industries received approximately 60% effective protection in their home market, with substantial variation from highly protected industries such as transport equipment and relatively exposed processors of raw materials.

Korea's trade policy was marked not by the absence of protection in general, but by its absence in those industries that had strong export potential and could best respond to export incentives. In other words, protection of the domestic market was high in industries in which Korea did not face strong export prospects, and it was low in industries in which Korean products were internationally competitive. Thus, while Korean policy did offer protection to the domestic markets of industries producing importables, it offered little incentive for industries producing exportables to keep their output at home.

Table 2.2: EFFECTIVE RATE OF PROTECTION, ALL SALES
(% of value added)

Sector	1968	1978	1982
Agriculture	18.5	57.1	74.3
Mining	4	-1.5	-1.7
Manufacturing	-1.4	31.7	28.2
Food	-18.2	-44	-48.4
Beverages, tobacco	-19.3	33.4	15
Construction materials	-11.5	11.8	51.1
Intermediate I	-25.5	37.6	61.9
Intermediate II	26.1	20.6	39.6
Nondurable construction	-10.5	67.4	42.4
Durable construction	64.4	242.9	52.5
Machinery	44.2	44.2	31.3
Transport equipment	163.5	326.6	123.9

Source: S. Young (1984).

Table 2.3: EFFECTIVE PROTECTION OF EXPORT SALES MINUS EFFECTIVE PROTECTION OF DOMESTIC SALES IN DIFFERENT TYPES OF SECTORS
(% of world value added)

Countries	Export	Export & import	Import subst.	Non-traded	All
Argentina	-5	-144	-189	-147	-38
Colombia	-2	-15	-50	46	-11
Israel	-51	-72	-91	-33	-60
Korea	35	-44	-52	3	-1
Singapore	-2	-11	-18	-3	-7
Major Competitor Economy	12	-1	5	-1	14

Source: Balassa *et al.*, (1982).

Financing

2.17 Support for exports was pervasively channeled through the state-controlled banking system. Government objectives were implemented through policy loans--bank loans explicitly earmarked for particular activities or industries, and lent passively by banks at interest rates below those charged for general lending purposes. Rates charged for export activities were particularly low. Special funds, administered by commercial banks, were established for export promotion (1959 and 1964) and for financing the inputs of export industries (1961). Following explicit priorities of Government, banks increasingly used export performance as the criterion of creditworthiness. Access to bank credit was extremely important, since the bank lending rate was substantially below the cost of borrowing in the alternative curb market, with the average spreads being approximately 22-25% in the 1963-1973 period (see Chapter 4 for details).

2.18 Extensive government control over financial and other resources is not unique to Korea, but the way control was exercised is noteworthy. The Government's overriding interest in export development was clearly communicated to the principal agents of the economy. Institutions such as trade promotion meetings, industry and firm-level export targets, close surveillance of export performance, and special awards for export achievements are legendary features of the Korea's export promotion system as it emerged in the 1960s. These instruments, backed by high-priority lending to exporters, gave trade performance extraordinary visibility and doubtless helped to focus the efforts of all economic institutions--firms, banks, and the bureaucracy--on the implementation of the outward-looking strategy. What made the system work was the government's commitment to exports and its ability to act decisively.

Rewards of Outward Orientation

2.19 The challenge of the 1960s was to establish a pattern of progress in an economy saddled by a history of foreign domination and extensive economic controls. The ensuing reforms combined an innovative mix of exchange rationalization, export promotion, and import liberalization. Above all, they signalled a major government commitment to export-oriented development. This strategy was rewarded with extraordinary export growth. The policy mix can be best described as offering broadly offsetting, rather than neutral, incentives for import substitution and exports. Indeed, in Korea even import protection served export promotion objectives by allowing firms to earn domestic rents if they exported a portion of their output. The level of intervention was high, but on detailed analysis, appears to have provided export incentives that were stronger than is normally the case in even neutral economies.

2.20 Throughout the 1960s, reforms aimed at greater neutrality--exemptions from tariffs and import controls--were coupled with positive interventions favoring exporters. Added together, these various subsidies apparently offset the protection afforded to domestic markets by high tariff and non-tariff barriers. The Government's focus on trade performance also explains why the discretionary components of the system achieved excellent results. Exporting was identified as the criterion of resource allocation, and the performance of firms, banks, and the bureaucracy was closely monitored with this target in mind. The system was decidedly interventionist, but it resolved the ambiguities that create corruption and waste in other settings in favor of a criterion closely related to economic efficiency.

2.21 As of the late 1960s, the principal features of the trade regime were the following: (i) moderate overall protection of domestic markets, offset by special subsidies to exports; (ii) approximately world market pricing of inputs and outputs across different export products; (iii) high protection of the domestic market in industries with poor export prospects; (iv) high protection of final consumer goods, modest protection of industrial raw materials and capital goods. Overall, protection did not significantly distort either the general level or the interindustry pattern of export incentives. Consequently, the emerging export structure reflected comparative advantage more closely than is generally the case in protected economies.

2.22 Korea's outward-looking trade strategy contributed significantly to the overall expansion of the economy. Very crude estimates, based mainly on national accounts data, show that export growth accounted for less than 10% of real GNP growth before 1960 (see Table 2.4). Its contribution rose during the 1960s, reaching over 20% in the first half of the 1970s. By the latter half of that decade, about a third of Korean growth could be attributed to the expansion of exports. The role of import substitution became progressively smaller; starting from small positive contributions in the early 1960s the effect of import substitution became negative, a situation which persisted throughout the 1970s, although it was substantially diminished during the HCI drive in the mid-1970s. The effect of changes in Korea's competitive position relative to the rest of the world (the sum of the export growth and import substitution effects) was mostly positive through the last 25 years, and has grown more significant with time.

Table 2.4: SOURCES OF GROWTH
(% of GNP growth)

Source/period	1955-60	1960-65	1965-70	1970-75	1975-80
Consumption	97.6	69.9	63.4	60.5	56.9
Government consumption	5.4	4.0	9.8	8.7	10.2
Investment	-5.6	22.7	53.4	30.1	40.5
Export growth	9.6	12.8	13.9	26.2	34.9
Import substitution	1.9	1.4	-18.2	-1.6	-11.8

Source: World Bank estimates.

2.23 Korean exports also contributed to growth by reducing the economy's capital/output ratio. Subject to a given level of savings, Korea's labor-using trade pattern effectively raised the ceiling on the potential rate of economic growth. The labor-intensive character of exports in turn, can be traced to the neutrality of export incentives.[8/] Exporting activities also contributed to the growth of Korea's technological capabilities.[9/] This reflects the positive effects of foreign competitive pressure on productivity and the fact that trade creates new channels for learning about and acquiring foreign technologies.[10/]

2.24 Korea's spectacular export performance during the period led to increasing confidence in the Government's ability to initiate and direct national development strategy. Certainly the takeoff period demonstrated that a favorable macroeconomic framework, combined with aggressive export-promoting intervention could lead to rapid growth. Buoyed by its past success, the Government next turned to more direct efforts to accelerate structural change, and plunged into the battle to promote heavy industry with an impressive arsenal of industrial policy instruments.

C. The HCI Drive: 1973-1979

Industrial Objectives

2.25 Despite intervention, there had been surprisingly little sectoral bias in Korea's development strategy prior to the early 1970s. Although special laws promoting machinery and shipbuilding were adopted in 1967, and basic materials and intermediate goods were frequently mentioned as an objective, the First and Second Five-Year Plans indentified labor-intensive exports as a high priority and the export imperative generally dominated government policy. The shift from general export promotion to a sectoral development strategy, focused on heavy and chemical industries (HCIs), was announced in 1973 by the late President Park. It represented a major change in policy in favor of specific industrial targets and a wide-ranging commitment by Government to using trade and financial policies to steer resources to the HCI sector.

2.26 The change in strategy had both political and economic roots. The opening of US relations with China and the fear of a possible withdrawal of American troops prompted the Government to seek an industrial base for an

---

8/ Exports required 25% less capital per worker than domestic demand (Westphal and Kim, 1977), and accounted for more than 50% of the growth in employment in the late 1960s (Watanabe, 1982).

9/ Using macroeconomic data, including data from the Korean experience, Nishimizu and Page (1985) find a clear correlation between technical progress and the degree of outward orientation.

10/ The positive effects of trade on productivity are confirmed by the microeconomic studies by Rhee, Dahlman, and Westphal (1985).

independent defense effort. On the economic side, the objective of "deepening the industrial structure" was seen as a logical response to the rapidly rising Korean wage rate and increased global competition in some of Korea's traditional export industries. The potential entry of China in world markets appeared to accelerate these trends. At the same time, Japanese penetration of global steel, electronics, and automobile markets provided a timely and successful model.

2.27 The oil shock of 1973-74 came after the HCI drive began, although some analysts suggest that it provided additional motivation for HCI policies. By demonstrating the riskiness of international commerce, it is argued, the oil shocks justified greater attention to import substitution. These arguments do not hold up well, since the HCI drive implied, in some cases, greater dependence on more concentrated world markets than the traditional export strategy. Several of the industries promoted in the HCI drive had large economies of scale, and efficient production implied capacities well beyond the scale of the domestic market.

2.28 In any case, the Government believed that the industrial structure needed "upgrading," and that the new directions required large-scale risky investments which would not be undertaken by private firms without decisive government leadership. Indeed, given the Government's control over industrial finance, no significant change in economic structure could have occurred without its consent. The targeted industries included naphta cracking, steel, metal products, shipbuilding, machinery, and automobile production. By 1980, these industries were to capture a substantially larger share of value added as well as account for more than 50% of Korean exports. However, the effort proved to be extremely costly.

Intervention Through Industrial Finance

2.29 The promotion of the HCI sector was supported by a broad range of policy instruments, including import protection and fiscal preferences. But the intervention which mattered most, and had the greatest impact on industrial incentives and structure, was the allocation of credit. Government relied heavily on its control of the entire credit system and provided "strategic" industries preferential access at substantially subsidized rates. The potential for subsidization was great due to the complicated system of interest rate ceilings that governed the financial system throughout the 1970s. The high interest rate policy of the late 1960s was discontinued in 1972 and lower interest rate ceilings ensued. Real bank interest rates were negative throughout most of the 1970s and created severe excess demand for bank credit. The differential between bank rates and those charged in the informal credit markets represented a substantial discount for industries eligible for credit from the government-controlled banks.

2.30 The use of policy loans was pervasive in the HCI context; for example, even at the end of 1981 there were reported to be 221 types of policy loans extant among the total 298 types of bank loans. In the 1970s, between 43% and 50% of total lending went for "preferential finance." By source, policy loans can be classified into those financed by Government, either directly or through Bank of Korea rediscounting on the National Investment

Fund (NIF) and those funded by banks. In terms of availability, policy loans involved either automatic access for designated activities (such as export loans or loans for equipment of an export industry) or qualified access (such as lending to selectively approved borrowers in a designated category, e.g., small and medium industry). While it is impossible to assess the full role of policy loans in the credit system,[11/] lending by the NIF and deposit-money banks (DMBs) provide good insight into how directed credits were used to underpin industrial objectives.

2.31 The National Investment Fund (NIF), established in 1974, lent as much as two thirds of its portfolio to HCI projects. Whereas the NIF was the most visible and most clearly directed financial support for specific industries, it only constituted between 3.0-4.5% of total domestic credit during this period. Therefore, its real impact on credit allocations stemmed from its "announcement effect" on bank lending practices. Using a conservative measure, namely the proportion of total DMB loans classified as policy (including export) loans, it is estimated that roughly one third of Bank lending, the predominant source of formal domestic credit, went into "policy" loans.[12/] Strategic industries, such as chemicals, basic metals, and fabricated metal products and equipment, received favored access to other bank lending as well, as compared to traditional industries producing either for the domestic market, such as food and beverages, or for export, such as textiles.[13/] Thus, the share of credit allocated to the three priority sectors virtually doubled from approximately one third of total DMB lending in 1973-74 to about 60% in 1975-77. At the same time, light manufacturing industry's access plummetted, as seen in Table 2.5.

2.32 Supported by NIF and other policy loans, capital formation accelerated sharply. Between 1976 and 1978, for example, investment grew by 27% p.a., but its primary contribution was to capacity in the HCIs. According to the KDB investment survey, the HCI sector claimed almost four fifths of manufacturing investment between 1977 and 1979. As a consequence, nearly all

---

11/ This is particularly true in the case of bank loans because Government clearly gave the nod to certain industries and banks were clearly aware of those signals.

12/ At the height of the HCI period, say 1977, W 1.6 trillion of the deposit-money banks' total credit availability of W 4.7 trillion was officially designated as policy loans. Adding the credit made available through specialized government banks (i.e., KDB and KEXIM), which was in many respects essentially analogously controlled, a total of 45% of total domestic credit of the banking system can be said to have been engineered in direct support of industrial policy objectives.

13/ See Chapter 5, Volume II (Table 5.13) for a calculation of the implicit subsidy awarded to officially sanctioned borrowers, measured either as the difference between the general bank lending rate and that for either policy or export loans or as the difference between that lending rate and the curb market rate.

Table 2.5: INCREMENTAL CREDIT ALLOCATION BY THE BANKING SECTOR
(%)

Manufacturing	1973	1974	1975	1976	1977	1978	1979	1980	1981	1982	1983	1984	1985
Food & beverages (1)	8.93	14.50	8.85 (23.1)	3.46 (21.4)	3.20 (20.8)	2.78 (19.5)	2.52 (19.2)	3.43 (19.8)	6.04 (20.7)	7.15 (21.4)	8.71 (20.8)	2.87	9.12
Textile & apparels (2)	42.23	34.78	12.87 (20.9)	34.94 (20.9)	26.59 (18.4)	23.37 (18.8)	18.54 (17.9)	17.40 (17.1)	20.13 (17.7)	10.86 (17.0)	13.98 (17.6)	26.24	14.51
Wood & furniture (3)	7.14	7.62	-8.17 (2.1)	1.20 (1.9)	1.43 (1.9)	4.88 (1.8)	2.84 (1.7)	2.88 (1.3)	4.93 (1.2)	4.83 (1.2)	4.74 (1.1)	3.09	3.35
Paper & printing (4)	4.84	5.60	2.29 (4.4)	3.91 (4.1)	2.50 (4.2)	4.63 (3.7)	6.16 (3.4)	5.99 (3.7)	5.21 (3.5)	4.01 (3.3)	7.31 (3.6)	6.09	5.29
Chem, petr. & coal (5)	7.58	11.64	25.26 (19.8)	17.29 (20.2)	15.98 (21.0)	16.09 (19.7)	14.06 (20.1)	19.61 (22.9)	16.06 (22.4)	15.25 (22.2)	15.63 (22.4)	12.74	16.21
Nonmetallic min prod. (6)	3.25	4.23	8.13 (4.7)	2.05 (3.9)	4.68 (4.5)	7.23 (4.2)	5.48 (4.9)	5.61 (5.0)	6.21 (4.6)	2.01 (4.7)	4.65 (5.1)	3.52	8.76
Basic metal (7)	12.29	6.97	8.91 (2.9)	18.89 (3.8)	12.31 (4.4)	12.20 (5.2)	14.61 (7.0)	10.35 (6.5)	16.30 (7.4)	17.27 (7.7)	4.71 (7.2)	7.19	8.11
Fab met prod & equip (8)	15.74	13.60	31.58 (19.3)	19.74 (21.1)	32.43 (22.0)	27.36 (24.9)	29.70 (24.0)	29.80 (21.2)	20.15 (20.5)	35.86 (20.3)	38.00 (20.6)	36.33	39.04
Other manufacturing (9)	-0.99	1.06	10.29 (2.8)	-1.48 (2.7)	0.88 (2.8)	1.47 (2.2)	6.09 (1.8)	4.93 (2.5)	4.96 (2.0)	2.78 (2.2)	2.28 (1.6)	1.92	-4.38
Heavy industry	35.60	32.21	65.75 (42.0)	55.92 (45.1)	60.72 (47.4)	55.65 (49.8)	58.36 (51.1)	59.76 (50.6)	52.51 50.3)	68.38 (50.2)	58.34 (50.2)	56.26	63.36
Light industry	64.40	67.79	34.25 (58.0)	44.08 (54.9)	39.28 (52.6)	44.35 (50.2)	41.64 (48.9)	40.24 (49.4)	47.49 (49.7)	31.62 (49.8)	41.66 (49.8)	43.74	36.64
Total	100.00	100.00	100.00	100.00	100.00	100.00	100.00	100.00	100.00	100.00	100.00	100.00	100.00

Notes: (1) The figures are the share of net credit increase of DMBs and KDB. The figures in parenthesis are share of value added.
(2) HCI includes (5), (7), (8) and light industry includes (1), (2), (3), (4), (6) and (9).

Sources: BOK, Economic Statistics Yearbook and National Income Accounts.

of the investment projected by the Third Five-Year Plan for the 1977-1981 period in heavy industry had been completed by 1979, while the investment planned in other industries was less than half complete. Nevertheless, neglected industries like textiles managed to survive, and indeed carry the brunt of Korea's export performance during the second half of the 1970s, while many of the HCIs lumbered with excess capacity (see Chapter 1, Volume II).

2.33 There were clear disparities with respect to credit access. As a result, favored sectors not only had better access to capital but also faced much lower average borrowing costs, defined as the ratio of financial expenses to total borrowing.[14/] Figure 2.2 reflects the financial favoritism shown to certain sectors, particularly HCI in the mid-1970s. Panel A compares the average borrowing cost of heavy industry to that of light industry; a ratio of 1.0 approximately reflects neutrality in industrial finance.[15/] Beginning in 1974, the year of inception of the NIF and the year in which heavy industry began to claim a decidedly larger share of preferential loans (see Chapter 10, Table 5.7, Volume II) the gap in effective borrowing costs began to widen in favor of HCIs. Indeed, between 1975 and 1978, the height of the HCI push, the cost of borrowing averaged approximately 25% lower for heavy than light industry. This disparity began to recede over the 1979-80 period of reform and is now approaching neutrality.

2.34 A similar bias is evident in the ratio of the average borrowing costs of large versus small and medium firms, as seen in Panel B of Figure 2.2. Once again, after an abrupt change in 1979-80, the ratio is approaching neutrality, as Government tries to encourage lending to small-medium industries. By similar measures the preference accorded to export vis-a-vis domestic industries also faltered during the import-substituting era of the mid-1970s. Beginning in the 1979-80 period, the bias in favor of exports was reestablished and persists through to the present. Panel C shows that the years in which export industries received the greatest financial preference are essentially those in which HCIs were supported least.

Additional Incentives

2.35 HCIs were supported by a variety of other incentives as well. A new Tax Exemption and Reduction Control Law (1975) gave five-year tax holidays, investment tax credits, and accelerated depreciation to designated "key" industries.[16/] At the same time, other businesses faced higher taxes: for example, the commodity tax exemption previously available to all exporters was

---

14/ See Y.J. Cho (1985).

15/ If loans to different industries carry different exposures to risk, a borrowing cost ratio other than unity would obtain even in undistorted capital markets. In Korea, however, banks were not permitted to charge risk premia, and in the absence of preferential treatment all bank customers would have faced equal borrowing costs.

16/ See B.Y. Koo (1984) for a detailed description.

FIGURE 2.2: TRENDS IN INDUSTRIAL FINANCE
(AVERAGE BORROWING COST COMPARISONS)

A. HEAVY VERSUS LIGHT INDUSTRY

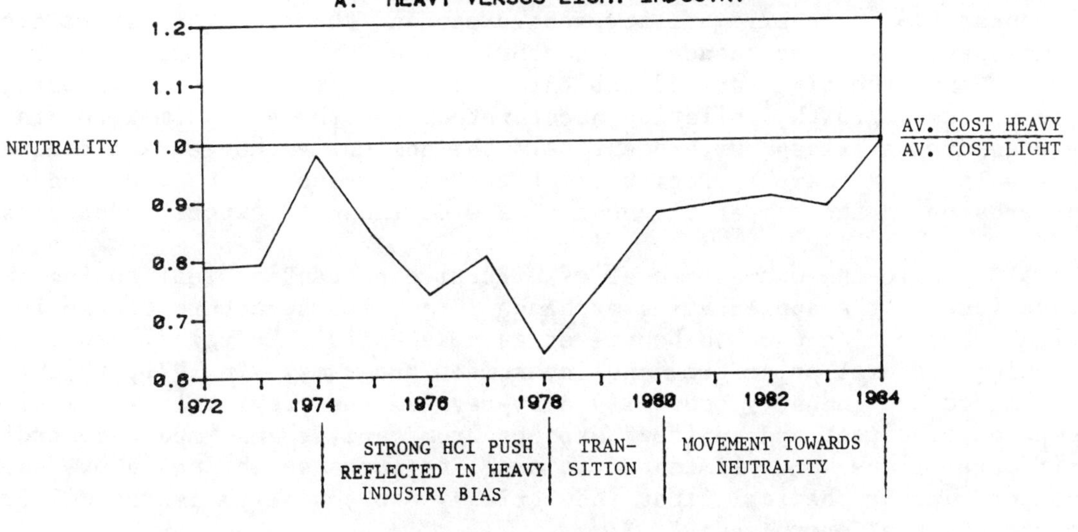

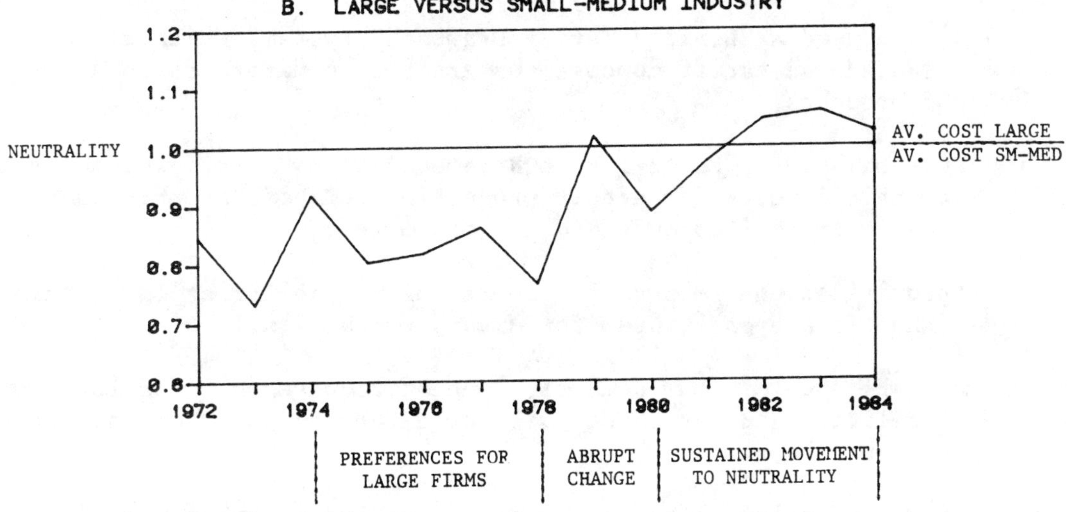

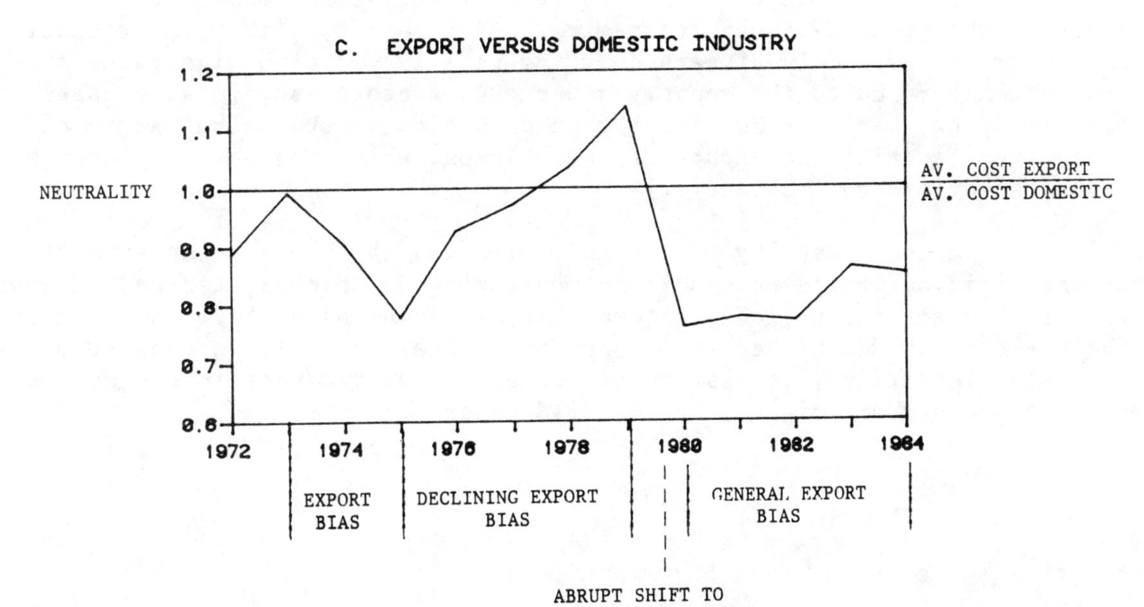

withdrawn. Important export preferences were reduced, including wastage allowances (1973), public utility subsidies, and the scope of export credit. The marcoeconomic consequences of the HCI drive further injured export industries. Since the high rate of capital formation was supported, in part, by rapid monetary growth, inflation accelerated. At the same time, to minimize the pressure on prices, Government held the nominal exchange rate constant. The result was a sharp appreciation of the real value of the won, and significant erosion in the competitiveness of a wide range of export industries.

2.36 While the overall level of effective protection fell during this period (due to the appreciating exchange rate), its structure became less benign. Protection came to be viewed as a selective, "surgical" tool for advancing sectoral priorities and investment programs. In 1971, the Ministry of Commerce and Industry (now MTI) also revised the previously automatic exemption of direct and indirect exports from tariffs and import controls. Tariff exemptions were eliminated in order to increase the sales of domestic producers in the chemical fiber industries. Subsequently, as the HCI drive progressed, the Government:

(a) introduced a "Limited Tariff Drawback" system, which eliminated or reduced the tariff concessions granted to exporters on a wide range of products;

(b) established a "Prior Import Recommendations" list, which permitted certain imports for export production only if the price advantage of the imported item exceeded a minimum level;

(c) prohibited the import of selected plant facilities in certain industries even if used for export production.

(d) established minimum domestic content requirements for large plant facilities and for those built with foreign loans or foreign currency loans.

These measures undermined one of the most effective and most durable components of the previous incentive regime. As a result, effective protection rates came to be widely dispersed by the late 1970s, with high rates of protection offered to the economy's targeted sectors and, in some cases, negative rates facing traditionally important exporters. The margin of incentives favoring the export of exportables, which was so important in the 1960s, deteriorated substantially.

2.37 Another casualty of the HCI drive was the close balance between private initiative and bureaucratic monitoring which characterized intervention in the earlier period. Interactive decision-making between business and the bureaucracy diminished in proportion to the overwhelming support provided for "key" industries; it was more important to participate in the HCI program than to produce efficiently or to build export markets.

2.38 Successes and Failures. It is now widely accepted in Korea that the HCI drive was overambitious and resulted in serious misallocation of resources. Nevertheless, in evaluating the policy, it must be observed that many of the goals of the policy were in fact achieved. Exports of heavy and chemical industries did not quite reach the target of 50% of all exports by 1980, but exceeded the target only a few years later and reached 56% in 1983. The value added structure of the economy also shifted sharply toward these industries (see Chapter 1, Volume II). Measuring success and failure is a normative exercise of some magnitude and not inconsequential difficulty. In a comprehensive, dynamic perspective it is difficult to demonstrate that an alternative policy would have worked better. Nevertheless, there is evidence that the HCI push produced, at best, mixed results, and that some of Korea's recent successes in heavy industry could have been achieved at lesser cost.

2.39 Industry-Specific Results. Industry-by-industry analysis of the HCI sector suggests major technical achievements. POSCO, the integrated iron and steel firm, is arguably the world's most efficient producer of steel. The firm has continued to operate near capacity despite a major slump in world steel markets and has recently concluded a long-term trade arrangement with U.S. Steel. Under this arrangement, which will greatly reduce the risk of protectionism in the American market, POSCO will supply both unfinished steel and technology to the US company. It should be noted that, while the industry has become technologically efficient, it was developed with highly subsidized capital and thus may not have been an economically efficient investment in and of itself, although it has provided benefits to the economy in a dynamic sense.

2.40 Consumer electronics has also achieved a strong international competitive position and is one of Korea's major growth sectors, although some of the most successful branches of the industry involve assembly of imported components and small-scale manufacturing--neither of which was particularly targeted during the HCI drive. As described in detail in the case study on electronics (see Volume II), Korea is investing heavily now in the high-technology area and is hoping to capture a significant share of the world's semiconductor market. This strategy involves major risks and high-stakes investments, which have so far been absorbed by the private sector in a sharp departure from past practice.

2.41 Other industries became competitive but faced unusually adverse changes in world markets. Shipbuilding sharply raised its share of world markets and operated profitably during the boom of the late 1970s. Even though world capacity utilization began to decline near that time, Korea's competitive position seemed so strong that capacity was sharply expanded. The utilization of this new capacity proved disappointing; utilization rates in the 1980s appear to have fallen as orders declined. Despite attempts to adjust to worsening market conditions by diversifying into platforms and smaller ships, the industry now faces serious potential problems of unemployment and financial distress.

2.42 Automobiles present a reverse case of mixed results. Early experiences with expansion were disappointing; in the domestic and international turmoil of 1980, utilization is said to have fallen to about one third of capacity. As recently as 1982, KDI studies showed that the industry was not globally competitive: its value added was negative at world prices. Yet recently the industry has had some startling market successes and all three manufacturers are now planning to expand capacity significantly with some guarantee of foreign markets. The unanswerable question is when and on what terms these deals would have come without the high-cost, premature investments of the late 1970s.

2.43 The weakest results of HCI investments are seen in segments of the petrochemical and heavy machinery industries. Fertilizers, for example, are a prime example of overambitious investment in an import-intensive industry. Investment in capacity far exceeded domestic demand. Based on naphta instead of natural gas, the competitiveness of Korean fertilizer was highly vulnerable to oil price changes. As of 1980 Korean urea production costs were five times as high as US and Canadian costs, and ten times as high as Middle Eastern costs. Substantial capacity had to be scrapped. Other petrochemical facilities eventually reached reasonable operation rates, but apparently not without implicit subsidization by domestic consumers.[17/]

2.44 Heavy machinery investments, particularly in the Changwon machinery complex, sharply overestimated Korean and world demand for electrical generators and equipment for heavy industry, and failed, on the whole, to produce competitively priced products. Some of the complex's huge foundry shops are said to be operating at only a fraction of capacity. The industry underwent a series of reorganizations in the early 1980s and is concentrating on its more managable businesses, with some recent success. But despite joint ventures with foreign companies and strenuous efforts to "localize" domestic machinery orders, heavy machinery capacity is unlikely to be fully absorbed in the near future.

Overall Assessment

2.45 Microeconomic performance adds up to macroeconomic performance. Not only did the HCI drive result in substantial unusable capacity, but it also concentrated investment in the economy's most capital-intensive industries. This was in sharp contrast with the development strategy pursued in the 1960s. During the takeoff, the rising share of exports in the economy led to a sharp initial decline in the capital-output ratio and then kept this ratio below its 1955-1960 level until 1970-1975 (see Table 2.6). These changes underscore the significance of changes in the economy's output mix, since the overall capital stock substantially increased during the same period. The capital-output ratio took a sharp turn upward with the HCI drive. A comparison of Korean capital-output ratios with similar measures for other newly industrializing economics shows that Korean ICORs in the 1960s were extraordinarily low (see Table 2.7). The upturn of the 1970s moved the Korean ICOR closer to those of

---

17/ Rhee (1985) shows that the prices charged to export users for petrochemicals range from 50% to 85% of the prices charged to domestic users; the prices charged to exporters presumably reflect prices charged on world markets.

Table 2.6: INCREMENTAL CAPITAL-OUTPUT RATIOS
(gross investment/output change)

Source/period	1955-60	1960-65	1965-70	1970-75	1975-80
Gross Investment over Period	1,257	1,657	4,993	10,607	21,266
Change in GDP	425	1,038	2,459	3,637	4,390
ICOR	2.96	1.60	2.03	2.92	4.84

Source: World Bank estimates.

Table 2.7: INTERNATIONAL COMPARISON OF ICORs
(gross investment/output change)

Country	1960-70	1970-80
Agrentina	4.6	9.0
Brazil	3.5	3.2
Greece	3.5	5.5
Hong Kong	2.4	2.2
Korea	1.8	3.6
Philippines	3.9	4.3
Portugal	3.3	3.7
Singapore	2.5	4.0
Thailand	2.7	3.5
Yugoslavia	5.5	5.4
Average	3.4	4.4

Source: World Bank estimates.

other economies; in the first half of the 1970s Korea was still near the bottom of the distribution, but by the latter half of the decade it had moved well above the average.

2.46 The structural transformations effected by the HCI drive were, on the whole, consistent with emerging changes in Korean comparative advantage, but occurred too rapidly and at excessive cost. It is true that in some industries the strategy worked, and that in many others disappointments were due to external causes. But the great variance in HCI outcomes cannot be separated from the *ex ante* risk assumed by the HCI program. Without the virtually unlimited government support that was offered to HCI investments, no private agent would have been willing to bear the obvious risks. These risks paid off in some cases, but on average produced low returns. (See Chapter 5, Volume II on industrial financing and performance.)

2.47 More fundamentally, the HCI program substituted bureaucratic judgment for market tests and absorbed too much of the economy's resources. In contrast to previous years, investment was not sufficiently conditioned on the test of export performance. In shipbuilding and other industries the investment program continued beyond the time when market signals turned negative. This was partly the result of long gestation periods, but it also reflected the neglect of the previously operating "feedback" mechanisms--between business and Government, exports and credit. The excesses of the HCI episode came home to roost with the macroeconomic crisis of 1980, precipitated by the second oil price increase, but clearly exacerbated by the import-substitution policies of the 1970s.

D. Cautious Liberalization: 1979-Onwards

The Shift in Policy

2.48 Korea faced formidable problems in 1979: the structural and macroeconomic imbalances created by the HCI drive were aggravated by the second oil crisis, and the assassination of President Park added great political uncertainty. The new Government simultaneously had to stabilize the macroeconomy, solve mounting financial problems in major industries, and establish new directions for industrial policy. It made progress along all three directions in a very short period of time.

2.49 Macroeconomic stabilization was the first priority and classic adjustment policies were put in place with assistance from the IMF and World Bank structural adjustment lending.[18/] While the macro shock which hit Korea was clearly external in origin, the industrial policies of the 1970s had reduced the economy's resiliency; inflation was already high, reflecting an overheated economy being pushed through rapid change; the exchange rate was overvalued; capacity utilization in the heavy industrial sector was low;[19/] and exports were faltering. Devaluation and a shift in credit allocation policies were quietly adopted. And some attempt was made to enforce more conservative financial behavior by the corporate sector.[20/]

18/ These policies and their effectiveness are adequately described elsewhere; see, for example, Y.C. Park (1985) and a forthcoming study by the KDI and World Bank on stablization.

19/ Metal products, machinery and equipment exhibited capacity utilization rates of about 60% during the late 1970s, for example, prior to the 1980 recession. Chemical products, the HCI industry operating at the highest capacity rates never recoverd from the 1979-80 shock and is still limping along. See Annex Table A4.2, Volume II.

20/ Government regulations attempted to constrain new bank credit to highly leveraged firms. It also signalled its reduced commitment as a risk partner of the corporate sector by not preventing the bankruptcies of several big firms.

2.50 More important in a longer perspective was the policy shift toward greater industrial neutrality, which was clearly articulated in the Fifth Five-Year Plan, written in 1979. The proximate cause for the shift in strategy was mounting evidence of financial losses and structural distortions caused by the HCI drive. Too large a share of the economy's resources, had been gambled on the HCI sectors, and the crowding out of traditional industries had clear negative effects. It was now time to redress the investment imbalance, increase efficiency, and lay the groundwork for future competitiveness by focussing on technology policy. Government also recognized that the complexity of the economy began to exceed its management capacities. The Plan's emphasis on indicative planning and a greater role for the market was eventually manifested in a range of financial and import liberalization programs.

2.51 With respect to the framework of industrial policies, Government:

- began to reverse its past preferences toward large, heavy industry firms by reserving credit for small and medium firms;

- reduced its role in specific credit allocation decisions, and abruptly terminated policies which awarded the HCI sector large-scale preferences (see Figure 2.2).

In the area of industrial finance, Government:

- sold the commercial banks to private shareholders, although it continued to exercise significant influence over banking decisions;

- established new financial institutions and permitted some growth in the international activities of domestic banks and the domestic activities of foreign banks;

- increased real interest rates, reducing the gap between the organized and unorganized sectors of the financial market;

- eliminated interest rate subsidies for particular borrowers, reduced the size of special funds, and more generally scaled back the role of policy targeting in lending decisions.

Finally, with respect to protection, Government:

- committed itself to increasing the liberalization ratio, from about 69% in 1980 to 95% in 1988, according to an "advance notice" schedule;

- revised the customs law in order to reduce product-to-product variations in protection;

- lowered the average legal tariff rate by approximately one third, and committed itself to further reductions between 1984 and 1988;

- somewhat liberalized the system of controls over foreign direct investment.

The details of these measures are analyzed more fully in subsequent chapters of this Report, but even a short summary illustrates the contrast between the government's role in the 1980s and its earlier approach.

2.52 The general thrust toward neutrality notwithstanding, Government has continued to take an active role in several areas of policy. Intervention since 1979 has focused on three areas: restructuring of distressed industries, support for the development of technology, and promotion of competition. An active role in these "functional" areas can be regarded as consistent with the liberalization effort, at least to the extent that it can be rigorously justified on the basis of market imperfections.

2.53 Restructuring operations have become frequent in the wake of sharp reversals in world markets and the overambitious investment programs of the 1970s. The firms and industries involved were large and highly leveraged, with their loans representing a significant share of commercial bank assets. Restructuring plans have involved mergers, capacity reduction programs, as well as general support to commercial banks. For example, Government:

- restructured the machinery industry by merging major heavy electrical equipment producers (Hyosung, Kolon and Ssangyong) in 1980 and assigned them monopoly status;

- nationalized Hyundai International's Changwon complex (renamed KHIC in 1980 under KEPCO management) and ultimately split it into components under the management of Samsung, Lucky-Goldstar and KEPCO (1983);

- merged and reduced capacity in the fertilizer industry (1982);

- initiated a shipping industry rationalization plan which reduced the number of companies from about 60 to about 15, combined lines of business, and encouraged scrapping of capacity (see case study on shipping in Volume II of this Report).

2.54 It is clear from these cases that Government has bypassed competitive solutions in most of its restructuring operations. Reluctance to permit market forces to guide adjustment on the surface to be seems on the surface to be inconsistent with the policy emphasis on liberalization. Nevertheless, there are justifications for some interventions in the Korean context. First, financial distress was so widespread that it threatened the viability of commercial banks as a group. Failure of major groups of commercial banks would have undermined confidence in Korean finance with serious reprecussions for, among other things, access to foreign capital. Second, Korean banks had little experience with "workouts" of financially distressed firms. Third, since nonfinancial firms are generally highly leveraged, they could not be counted on to finance mergers or buyouts of other, troubled firms. Fourth, the investment programs that generated financial distress had been encouraged, in many cases, by past government policy. Letting firms fend for themselves now would have increased private perceptions of investment risk and undermined Government's ability to implement policy in the future.

2.55 While there are justification for intervention, the restructuring exercises of the early 1980s have also created problems of "moral hazard." An active government role reduces incentives for tough, private adjustment programs and encourages firms and industries to wait for public rescue. For example, some observers have charged that distressed companies, expecting government intervention, have chosen to borrow rather than scale back capacity and employment. These forces are strong in the oligopolistic settings typical of Korean enterprise: a weak firm may postpone adjustment until the "rescue," hoping that its share in the final merger or cartel will represent an improvement over scaling down or accepting a private merger proposal. These considerations are particularly relevant to rescue attemps structured along the Japanese model which tends to distribute cartel gains roughly in proportion with precartel market share. These issues are discussed further in Chapter 4, Volume II.

2.56 Intervention in technology promotion has stressed the establishment of institutions to train scientists and engineers and conduct basic and applied research. Under the Fifth Five Year Plan, national science and technology investment was to be increased from 0.9% of GNP in 1980 to 2% in 1986; the Sixth Plan aims for a 2.5% ratio, roughly equal to OECD spending, by 1990. The public budget (roughly 40% of Korean R&D spending) has supported general research and scientific training, as well as special research centers for energy and resources, machinery, electronics, telecommunications, chemicals, and tobacco. In addition, a National Project for Research and Development (1982) was established to fund public as well as public-private "joint" R&D projects in the high-technology fields of electronics, fine chemistry, and engineering. With the help of these programs, and new tax incentives under the Technology Development Promotion Act (enacted in 1973 and strengthened in 1981), private research and development expenditures expanded rapidly; for example, the number of private research centers doubled between 1982 and 1984. These issues are discussed further in Chapter 1, Volume II.

2.57 Finally, Government has shown growing interest in institutions that will enhance competition behavior. A Fair Trade Law (1981) has been adopted designed to guard against anticompetitive mergers as well as unfair advertising and restrictive trade practices. Most of the Office of Fair Trade's activities so far have involved domestic unfair trade practices and have been resolved through the voluntary cooperation of offending firms. Policy on mergers and cartels in the context of restructuring has been set by the high-level Industrial Policy Council and has not generally involved the Office of Fair Trade. Measures to reduce business concentration have been in effect since the early 1970s and received special attention in the Special Presidential Directives of 1974. Under these directives, firms were to issue public shares, as well as take steps to divest unrelated business and real estates. In general, these measures have not been particularly successful; the share of the conglomerates in GNP rose markedly in the late 1970, in part because the HCI drive emphasized scale-economy activities managed by larger firms.[21] In 1985 the Government restricted additional bank lending to the five largest groups to their share of outstanding loans in 1983, but then quickly lifted the restriction due increasing pressures on export performance. Concentration issues are discussed further in Chapter 1, Volume II.

Evidence of Increasing Neutrality

2.58 There are preliminary indications that the reorientation of Korean industrial policy begun in 1979 is taking root. Major financial distortions with respect to access to preferred sources of credit, as measured in Figure 2.2, have been reduced. Specifically, disparities in average borrowing costs between large and small firms and between heavy and light industry have narrowed substantially over the 1980-84 period. Also, perhaps to compensate for some lingering protection, export industries have again begun to obtain funds at lower cost than domestic industries, although even in this area there is a recent trend towards greater neutrality since 1982. While Government abolished the preferential lending rates of policy loans in June 1982 as part of the financial liberalization program, there still exists a difference between lending rates in the formal and informal sectors, and this constitutes an opportunity for the continued rationing of bank credit. But rapid growth in the amount of credit available through the nonbanking sector and direct credit market has served to reduce market segmentation and has tended to equalize borrowing costs to a greater extent across industries. These trends will be discussed in Chapter 4.

2.59 Converging trends can be also observed in the average returns to investment in different industries, as measured by the profits to assets ratio.[22/] Large and small and medium firms earned roughly equal rates of return in 1972-73. This parity disappeared during the decade, as performance by large firms faltered, but has reemerged in the 1982-84 period. A similar pattern emerges in the comparison of returns to capital in the heavy versus light manufacturing category. Here disparities existed over most of the period, but heavy industry's lag became especially severe in 1978-80, coinciding with HCI debacle, the oil price increase, and global slowdown. The interesting feature of all these measures of performance disparity is the trend towards greater parity (i.e., neutrality) during the reforms of the 1980s. (See Figure 2.3.)

2.60 Greater equality of credit access is undoubtedly important in explaining converging trends in industry profitability. For example, during the 1972-81 period, bank and foreign loans accounted for 41.4% of total assets of large firms and 32.1% of small and medium firms. These proportions shifted radically towards neutrality over the 1982-84 period--to 30.3% and 31.8%, respectively.[23/] These trends reflect Government's policy of increasing the

---

21/ Between 1977 and 1981, for example, the 30 largest Korean firms saw their sales share rise from 32% to 40% of mining and manufacturing. See section on concentration in Chapter 1, Volume II.

22/ See Chapter 5, Volume II for details.

23/ See Table 5.7 in Volume II of the Report.

FIGURE 2.3: TRENDS IN INDUSTRIAL PERFORMANCE
(Average Returns to Capital Measure*)

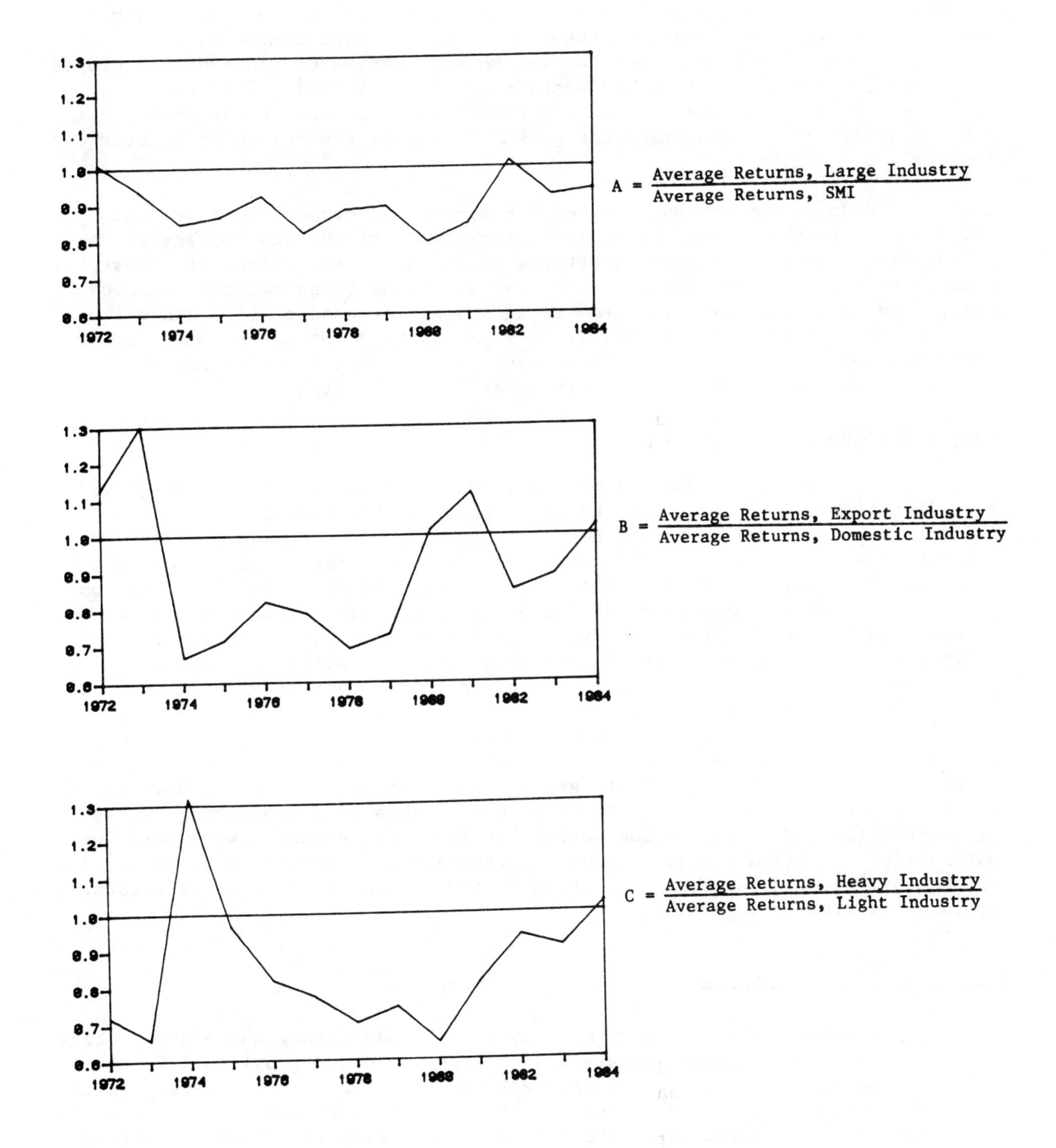

* Defined as financial expenses and profits as a fraction of assets.

Source: Bank Staff estimates.

share of bank lending that goes to the SMI sector, in an effort to reverse the distortionary effects of past credit policies. While the direction of change is helpful, large borrowers probably continue to have wider borrowing options, despite their high indebtedness, because the banks, and by extension the Government, became risk-sharing partners in the course of the 1970s and have yet to extricate themselves. At the same time, the more competitive allocation of credit is forcing reductions in the debt ratios of previously promoted manufacturing activities from the peaks reached in 1980-81. Debt equity ratios are still quite high, averaging around 3.5:1 in the recent years, but government efforts to encourage the growth of equity are beginning to bear fruit.[24/]

2.61 Despite concrete actions on the part of Government to reset the balance of incentives, and the general consistency of the new industrial orientation, there are lingering effects of the distorted system of industrial finance that characterized much of the 1970s. Among those dubious legacies are (i) an exceptionally highly leveraged industrial sector, (ii) generally low industrial profit ratios, and (iii) a pervasive dichotomy between performance and borrowing costs. For these reasons, industrial policy cannot proceed independently of domestic financial liberalization.

Evaluating the Reform Effort

2.62 The policy innovations of the 1980s are still too new to be judged for effectiveness. It is clear, however, that the combination of policies reflects aspects of an important new philosophy of economic management. The Government's earlier interventionist role precluded a rapid transition to "hands-off" management, both because of the inexperience of private decision-making institutions and because of the overhang of "real" problems such as widerspread financial distress. But Korea has embarked on establishing an infrastructure for greater market control of economic activity. The new policy regime is cautiously building reinforcing, decentralized institutions in several areas of policy. Government's decision-making powers are gradually being transferred to firms as sectoral investment priorities are abandoned, to capital markets as banks and financial institutions are liberalized, and to product markets as import controls are relaxed. At the same time, Government is strengthening the efficiency of firms and markets by establishing "rules" of competition and by supporting activities (such as research) with positive externalities. Given the Government's relatively good record in economic management, and the immaturity of private allocational mechanisms, a gradual approach is clearly defensible.

24/ The greatest gains have been made by export industries, whose debt ratios have fallen from approximately 5:1 in 1976-78 to about 3:1 in 1982-84. In general, the rise in real interest rates and the overall reduction in the thickness of the public sector "cushion" have forced firms to try and limit their indebtedness. These factors have also served to push firms into securing more equity capital than before, although the results so far are modest. See Table 10 of the Technical Appendix.

2.63 The view that substantial liberalization is under way is not universally shared, however, by all who watch the Korean experience with care. Some argue that Korean policy has changed in degree but not in direction, and that Government continues to wield great influence in industrial strategizing and implicitly in decision-making. The financial system may be undergoing liberalization, but it is still essentially credit-based and heavily subject to Government influence. Some analysts claim that Korea is not moving toward a stronger reliance on purely market forces, and that more broadly perhaps it shouldn't be, since, on the whole, intervention has produced good results.[25/] Others more modestly argue that it is premature to imagine a sea-change in the conduct of industrial policy, given that the actual reforms witnessed so far have involved mainly changes in the objectives pursued and the overtness of policy tools.[26/] In any event, Government is sending out revised, although not uniformly consistent, signals concerning its own preferred level of intervention.

2.64 The speed with which changes take hold should be judged within the context of Korea's legacy, embodied not only in objective economic conditions, as reviewed in previous sections, but also in the capabilities of allocative and decision-making institutions, and in the expectations of economic agents. Specifically, the Government's agressive industrial policy during the 1960s and 1970s has retarded the development of private institutions that share in industrial decision-making and risk-bearing. Equity markets are small; the financial decision-making experience of commercial banks and corporate finance department is thin. By contrast, Korea has strong institutions for business-government cooperation. High-level business and trade associations have been nurtured to direct and monitor the implementation of past policies. These institutions expect continued government direction, and frequently criticize the Government for not being activist enough. Firms and institutions that have become successful by working closely with the bureaucracy are reluctant to see the rules change. Consensus-building is clearly an enduring requirement for economic policy in Korea.

2.65 A complicating factor is that Korea is no longer simply developing new industries, but is also facing issues of industrial decline. Claims for government support--principally for credit access, low interest rates, and sometimes trade policy--are more balanced as between "infant" industries and industries in distress. In this new context, efficient adjustment requires more no's than was consistent with the facilitation of sectoral growth in the past: the priorities of "older" sectors and their economic interest groups are less and less likely to coincide with national economic interest. Vestiges of the former system of policy implementation may become lobbies against change. The Government's new industry consultative councils, modeled on the Japanese system, are supposed to help solve this problem: it is too early to tell how the approach will work, since in Japan it has meant

---

25/ See Wade (1985).

26/ See Westphal and Pack (1985).

government-led negotiations to spread losses, often with the consumer paying the bill. (See discussion on Japanese restructuring in Chapter 2, Volume II.)

2.66 For all these reasons, Government's effort to reduce its direct role in the economy will take time. Initially, economic agents will not readily accept that industrial policy has changed and will continue to look to Government for direction and support. During the transition, Government will be accused of creating uncertainty by abandoning its leadership role. Only with concrete experience will private agents recognize that Government's unwillingness to support an industry directly is not evidence of its lack of promise or profitability -- just as its past support did not always guarantee profits.

2.67 The fundamental lesson of Korean policy experience is that the economy is highly responsive to incentives. Strong response to incentives was demonstrated by both exports and imports with respect to changes in the real exchange rate and other trade policies, by investment with respect to the availability of credit, and by saving through the organized financial system with respect to the interest rate differential between bank deposit rates and the curb market. The responsiveness of the economy testifies to vigorous economic decision making in the private sector and this sensitivity to market signals should increase, *ceteris paribus*, as market imperfections are lessened. Its implications are positive for the effectiveness of any policy regime with allocationally correct incentives. The efficiency of resource allocation signals is the subject of the subsequent chapters on trade and financial liberalization and industrial policy.

CHAPTER 3: TRADE LIBERALIZATION

A. Background of Protection

3.01 The importance of the recent round of trade liberalization can best be understood in the context of Korean trade policy of the previous two decades. Out of a classically protected mold emerged a strong export-oriented trade regime in the 1960s, spurred on by a bevy of export incentives, including, inter alia, tariff exemptions, generous wastage allowances, preferential credit access, and sizeable tax advantages; at the same time, the level of protection provided to domestic producers remained generally quite high. Following the success of these outward-looking trade policies,[1] attempts began in the late 1960s to reduce the costs to society of highly protected import substituting industries. Buoyed by a stronger payments position, the government removed some quantitative restrictions and moved from a "positive" list of allowable imports to a tripartite system of prohibited, restricted and unrestricted imports in 1967. The reform movement was short-lived, however, and the ratio of restricted items, on the socalled "negative" list increased once again from 40% in 1967 to 49% in 1972, essentially where it remained during Korea's strong import-substitution phase, the HCI push of the later 1970s.

3.02 Although estimates of effective rates of protection (ERPs) are subject to error from many sources, it is clear from the numerous reported calculations that ERPs were dispersed, indicative of substantial protection was offered, especially to the transport, consumer durable, and machinery industries in the late 1960s. Thus, Korea's trade regime could legitimately be described as both outward-looking in the export side and restrictive on the import side. The nature of protection changed considerably in the 1970s as (i) industries affected by the heavy and chemical (HCI) push of the middle and late 1970s received additional protection; (ii) greater protection was offered to agriculture; and (iii) some heretofore less protected consumer goods also began to receive assistance. As a result, effective protection rates rose towards the end of the decade and all three ERP estimates for 1978 considerably exceed those for 1968 (see Table 3.1).

3.03 Protection in Korea has tended to rely on quantitative restrictions rather than tariffs, and neither tariff rates themselves nor actual tariff revenue collections show the marked change in the trade regime that accompanied the HCI episode. Industries targetted for rapid expansion in Korea's Third Five-Year Plan (1972-76) included petrochemicals, steel, machinery and shipbuilding, and these industries, in addition to receiving active positive incentives were also subject to domestic content regulations and more selective protection. The proportion of iron and steel products on the restricted

---

1/ See Kim (1982), Kim and Westphal (1977), Krueger (1976), Nam (1985), Young (1985) and Westphal (1978) for discussions of Korean trade policies and economic performance.

Table 3.1: EFFECTIVE PROTECTION COMPARISONS
(%)

	1968-70		1978			1982
	Kim & Westphal /a	Kim /b	Nam /c	Kim /d	Young & Yoo /e	Young & Yoo /f
Nominal protection /g	14.0	22.7	17.8	36.3	25.2	31.7
Effective protection /h Corden	9.0	21.1	24.1	28.3	34.1	38.4
Highest ERPs Corden method /i	Transport (83.2) Consumer durables (39.8) Machinery (29.5) Agriculture (17.9)	Transport (105.6) Machinery (72.8) Consumer durables (66.6) Construction (31.7)	Consumer durables (84.0) Transport (73.8) Agriculture (73.4) Machinery (33.2)	Agriculture (87.3) Transport (41.0) Mining (27.8) Inter. inputs II (26.9)	Consumer durables (119.4) Transport (108.8) Agriculture (54.5) Cons. nondurables (42.2)	Agriculture (70.6) Transport (60.4) Inter. input I (39.7) Cons. durables (36.0)

/a L. Westphal and K. S. Kim, "Industrial Policy and Development," World Bank Staff Working Paper No. 263 (August 1977).

/b Kim, Kwang Suk, "Long-Term Variation of Nominal and Effective Rates of Protection," Korea Development Institute (1982).

/c Nam, Chong Hyun, "Trade Policy and Economic Development in Korea," Discussion Paper No. 9, Korea University, mimeo (April 1985).

/d Kim, K. S., op. cit.

/e Young, Soogil and Yoo Jungho, "The Basic Role of Industrial Policy and a Reform Proposal for the Protection Regime in Korea," Korea Development Institute (December 1982).

/f Young and Yoo, op. cit (using 1978 input-output relations).

/g Nominal protection rate for all industries.

/h Effective protection rate for all industries.

/i The difference between Balassa and Corden measures is that the former treats nontraded goods as if they were traded and subject to zero tariff while the latter assumes that nontraded inputs are part of value added.

list jumped from 28% in 1967 to 75% in 1978 (on an items basis) and from 35% to 91% (on a value basis), for example. Similar patterns can be seen in chemicals and machinery.[2] One unique and important aspect of Korea's restrictive import regime was that it was not equivalent to absolute prohibition, nor did it include outlandish tariff levels. In fact, in 1978, at the height of the selectively protectionist phase, 58% of Korea's $14.6 billion in total imports were of restricted items, and indeed 75% of all manufactured imports (by value) were classified as restricted.[3] Thus, importation was carefully scrutinized and controlled by Government, but approvals were certainly given if determined to be in the national interest.

3.04 The sector offered the largest increase in protection during the 1970s and into the early 1980s has been agriculture. Clearly inefficient in terms of world prices but of considerable national sociopolitical importance, agriculture, in particular food grains and tobacco, was heavily shielded from external competition, a situation also prevalent in many industrialized countries. Protection offered to items like rice, wheat, grains and pulses exceeded 150% in 1982 according to KDI estimates. Increased protection was also accorded to consumer goods being produced for export, such as textiles. A plausible explanation for high protection in the domestic market rests with the need to produce rents on domestic sales in order to allow exporters to absorb thinner export profit margins. This approach, reminiscent of the Japanese experience, is consistent with Korea's trade policies throughout its modern history, namely, export promotion implicitly financed by domestic consumers.[4] In general, the protection pattern between 1978 and 1982, according to one estimate noted in Table 3.1, has increased somewhat, but the increase has been essentially in agriculture. Manufacturing protection, even for the most protected industries such as transport equipment, chemicals and technical machinery fell slightly, but was offset by increases in agriculture and processed foods.

3.05 This trend towards less pervasive industrial protection was the result of a shift in policies towards the end of the 1970s. Encouraged by a markedly improved balance of payments position, Government began in 1978 to both decontrol imports by adding items to the automatic approval list and

---

2/ Hong (1979, 1985) shows that as early as 1972 a number of chemical products were moved from automatic approval to the restricted list as new domestic production facilities came on stream. Targeted machinery industries were supported by increasing the domestic content ratio imposed on buyers of new machinery and by offering advantageous credit terms to buyers of domestic machines. The proportion of restricted items in machinery nearly doubled, from 34% to 61% between 1968 and 1978.

3/ The largest categories of restricted imports were machinery and raw materials.

4/ It can be argued that some of these biases were in essence implicit consumption taxes to promote export growth and help boost employment in export industries.

cutting tariff rates. An Import Liberalization Task Force was created and the advance notification system was instituted. Although further progress was slowed in the aftermath of the second oil price increase and the unrest following the President's assassination, the goals of increasing the pace of decontrol and both lowering tariffs levels, and reducing their variance were to be pursued with vigor in the 1980s. To understand the importance of the reform program, it is worth noting that there is no strong constituency for liberalization in Korea, in part because consumption goods, the items most heavily protected, account for only about 5% of the current import bundle, but also because there is a strong Korean sense of self-reliance. Liberalization per se is only seen as a useful process if it: (i) improves external competitiveness by reducing input costs of export industries or forces domestic producers to be more efficient; or (ii) is necessary to appease trade partners and ensure that export markets remain open to Korean products. Using these arguments, Korean policy-makers persuaded the public to support an aggressive liberalization program, which is scheduled to make consumer durables such as automobiles and computers importable without impediment by 1988.

B. The Current Round of Liberalization

Specifics of Reform

3.06 Trade liberalization in Korea has taken the form of import control relaxation and tariff reduction. The government began in the late 1970s to remove direct restrictions on imports and to move larger numbers of items to the socalled "automatic approval (AA) list." Automatic approval means that items are on the unrestricted rather either prohibited or restricted goods lists, a prerequisite administrative step for unencumbered importation. As of 1980, for example, 68.6% of imported goods categories were so designated. This proportion was raised to 74.7% in 1981, and as part of a major structural adjustment program, trade liberalization was aggressively pursued, the consequence being a steady rise in the "AA ratio" since then to a current level of 87.7% in 1985. The ultimate objective of this policy is to reach a liberalization ratio of 95.2% by 1988, as seen in Table 3.2. Industry-specific trends for the liberalization ratio are shown in Figure 3.1. At present, approximately three quarters of Korea's imports on a value basis are on the unrestricted importation list and this proportion is expected to rise in the final stages of the program.

3.07 The process of liberalization has been designed so as to minimize disruption while still promoting the desired effects of reducing distortions and opening the Korean market. The items subject to control are announced well in advance of the actual date of decontrol, so as to provide domestic industries with sufficient time to adjust to new external competition. It is reported that items initially given priority in the decontrol process were those which were fully competitive and those which strongly affected domestic price levels. As reported in Table 5.2, the liberalization ratios were quite dispersed at the beginning of the process, with only about 40% of electrical equipment on the unrestricted list but well in excess of 80% of chemicals freely importable. Since almost all industrial goods categories are expected to reach "liberalization ratios" of 95% or better by 1988 (viz., the major protected commodities will continue to be primary products), those industries

Table 3.2: IMPORT LIBERALIZATION PROGRAM
AUTOMATIC APPROVAL PROCESS

	Total items /a	Proportion of items subject to automatic approval							
		1981	1982	1983	1984	1985	1986	1987	1988
Primary products	1,386	68.5	70.6	73.5	75.8	78.2	79.7	80.1	80.5
Chemical goods	2,182	93.4	94.0	94.6	95.0	95.6	97.7	99.1	99.6
Steel and metal products	802	88.9	89.7	90.9	92.8	95.6	99.4	100.0	100.0
Machinery	1,414	64.2	65.5	69.2	78.0	83.0	89.4	93.3	100.0
Electrical and machinery appliance, electronics	495	40.9	46.1	51.3	62.4	73.8	87.0	95.6	100.0
Textile (including leather garments)	1,089	65.4	68.4	79.9	90.3	93.1	96.1	96.9	98.8
Others	547	71.2	75.7	80.6	82.1	82.8	85.7	88.2	88.2
(Subset A)*		-	(44.5)	(48.8)	(72.4)/b	(78.0)	(93.3)	(97.6)	(98.8)
Total (Items)	7,915	74.7	76.6	80.4	84.8	87.7	91.6	93.6	95.4
Total (Value)		64.8/c			72.2 /c				

* Subset A refers to those items on the monopoly list regulated by the Monopoly Commission, and defined as commodities whose sales exceed W 30 billion and either a single producer whose market share exceeds 50% or top three producers whose combined market shares exceed 70%.

/a As of July 1, 1984.

/b Based on 254 items, rather than the 354 used as the basis prior to 1984. The ratio based on the larger set of market-dominating items was 52.0% in 1984.

/c Estimate by Wontack Hong (1985).

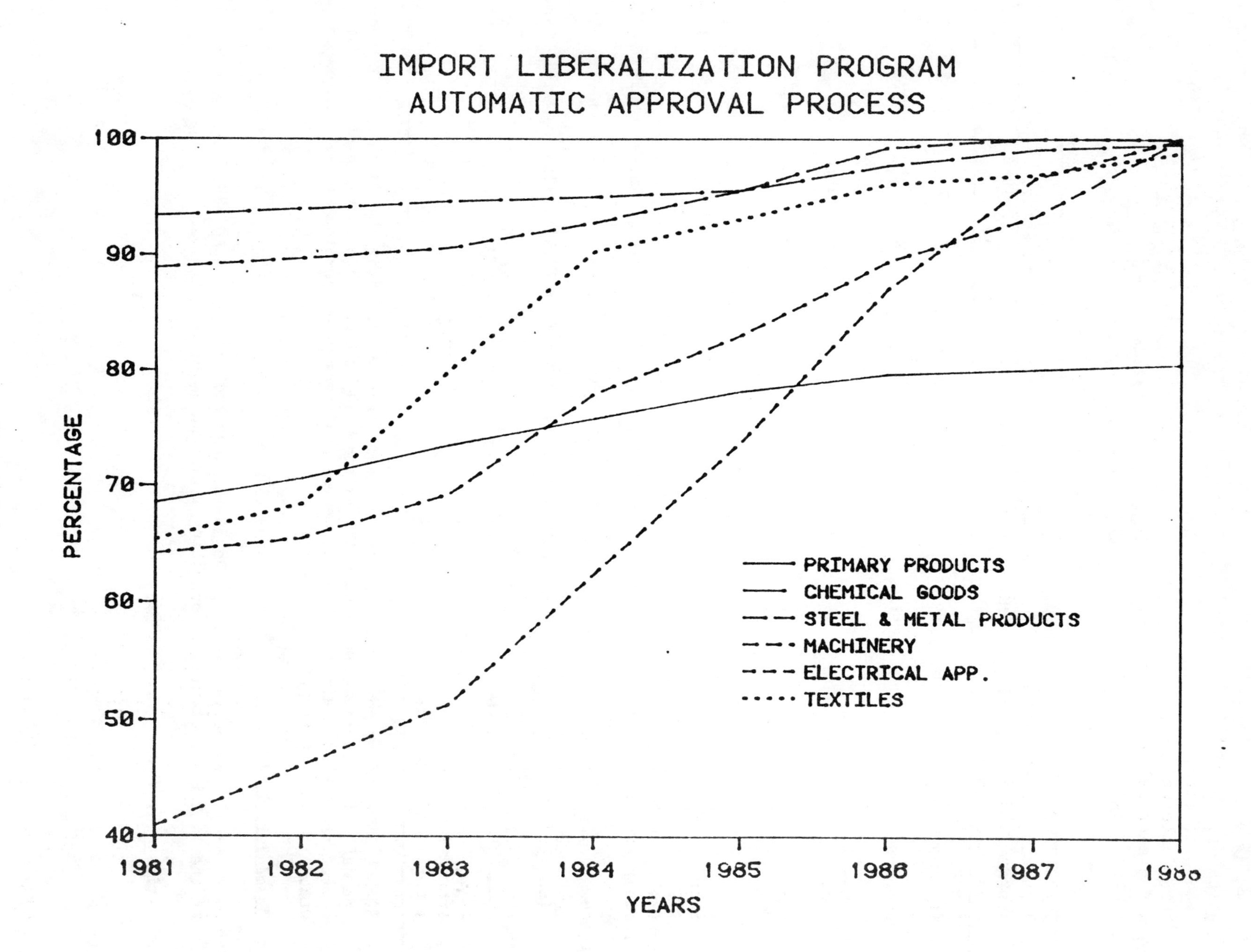

IMPORT LIBERALIZATION PROGRAM
AUTOMATIC APPROVAL PROCESS
PERCENTAGE
100
90
80
70
60
50
40
PRIMARY PRODUCTS
CHEMICAL GOODS
STEEL & METAL PRODUCTS
MACHINERY
ELECTRICAL APP.
TEXTILES
1981
1982
1983
1984
1985
1986
1987
YEARS

Figure 3.1

which were slowly liberalized in the early 1980s will have their liberalization pace accelerated in the 1986-88 period. In the next three years, therefore, external competition will most affect import categories in the machinery and electrical area, as well as the small subset of goods on the official monopoly list.[5/]

3.08 The number of new items to be moved from the restricted to the unrestricted list during the 1984-88 period is seen in Table 3.3. Essentially 357 items, mostly machinery and textiles, were added to the list in 1984 and another 233 items, including electrical equipment, machinery, and appliances were added in 1985. As a result, the number of restricted items was reduced from 1,560 in 1983 to 970 in the second half of 1985. Further liberalization is planned and published, so that it is expected that only 379 items will remain restricted at the end of the program in 1988. To date Government has stuck to its announced liberalization pace, despite clamors from some industrial quarters and difficulties in the conduct of trade diplomacy, maintaining that the efficiency gains to the economy continue to outweigh the adjustment costs borne by a few segments of society.

Analysis of Reform

3.09 In order to fully understand and assess the nature and scope of the trade liberalization program, attention must be focussed on the automatic approval concept *per se*. Granting AA may mean both more and less than meets the eye. It may mean less in the sense that a very large proportion of non-AA items are nonetheless imported by Korea, so that a restricted designation implies administrative control but in most cases, particularly for nonagricultural items, it does not pose an absolutely binding trade barrier. In that connection, it is instructive to note that in 1984 a total of $8.5 billion (27.8%) of Korean imports required some form of administrative approval because the commodity in question was not on the AA list. This included some $5.4 billion in transportation equipment, $1.5 billion in machinery, $1.3 billion of electrical items and almost $1.0 billion of chemicals.[6/] The automatic approval concept is imperfect and suggests more openness than it actually measures because items so designated may still be subject to some other possible administrative review procedures, such as the Special Law mechanism. Thus, the conventional "AA ratio" is biased upwards.

3.10 Special laws tend to impose health, safety and other public standards on import activities; however, as laudable as the ultimate goal may

---

5/ Major items scheduled for decontrol in 1986 include color TVs, buses, coffee, woolen fabrics, nylon carpets, diesel engines and electronic amplifiers. In 1987, items to be added to the automatic approval list include personal computers, razors, small cars, cranes, teleprinters, mini-buses, integrated circuits, dump truck and canned meats. In 1988, additional items include large cars, cameras, VCRs, vending machines, ball bearings, trailers, watch movements and crane-movers.

6/ See Hong (1985) for details.

Table 3.3: OPENING IMPORT MARKETS

	Restricted items /a	1983	1984	1985	1986	1987	1988 Restricted items
		(Items moved from restricted list to automatic approval list)					
Primary products	368	33	33	22	5	5	270
Chemical goods	117	10	13	45	30	10	9
Steel and metal products	74	16	23	30	5	-	-
Machinery	435	123	72	91	55	94	-
Electrical and machinery appliances and electronics	241	55	57	65	42	22	-
Textiles (including leather garments)	219	114	30	33	9	10	23
Others	106	6	5	16	14	-	65
(Subset A) /b	(10)	(21)	(14)	(39)	(11)	(3)	(3)
Total Restricted Items	1,560						367
Newly liberalized items		357	233	302	160	141	

	1982	1983	1984	1985	1986	1987	1988
Total liberalized items (1)	5,791	6,078 (6,355)/a	6,712	6,945	7,247	7,407	4,548
Total restricted items (2)	1,769	1,482 (1,560)/a	1,203	970	668	508	367
Total Items (3)	7,560	7,560 (7,915)/a	7,915	7,915	7,915	7,915	7,915
Automatic approval ratio (1) (3)	76.6	80.4	84.8	87.7	91.6	93.6	95.4

/a Adjustments made to conform to 7,915 items, the CCCN classification system initiated in July 1984.

/b Subset A refers to goods on the monopoly list. See definition of monopoly on Table 2.2.

possibly be, the net result of special law designation is to make importation potentially more difficult. Administratively, the Ministry concerned is obliged to announce in advance the conditions under which trade in a publicly restricted commodity will occur.[7] Clearly, greater transparency in the application of special law provisions is desirable and a reexamination of those provisions would be timely, inasmuch as a total of perhaps 1,874 items are currently potentially covered by one special law or another. Excluding fuels which are covered under the Petroleum Industry Act, an additional $5.3 billion of Korea's 1984 imports were subject to these provisions, primarily foodstuffs, raw materials, and chemicals. It is not known, of course, to what extent these administrative procedures lowered imports, i.e., when the relevant industry association or other body denied the importation of a commodity on the AA list.

3.11 The use of the liberalization ratio should therefore only be seen as a general yardstick of the intended openness of the economy, since even if items are on the automatic approval list, other trade measures, such as the surveillance system, the diversification system, and the aforementioned special laws may hinder their free importation. The surveillance system involves the designation of certain import commodities, which, while unrestricted, still require review by the appropriate industry association [8] because it is judged that their importation would either disrupt the domestic market or seriously damage domestic industry. As of 1984, there were 142 items covered by this system. The import diversification system aims to redress serious bilateral trade imbalances by attempting to discourage imports of selected items from particular countries. It is often used as a defensive device vis-a-vis trading partners whose own markets are clearly and artificially difficult to penetrate. As of 1984, 168 commodities were so designated.

3.12 There are risks in placing too great a reliance in the liberalization ratio. Since it tends to over estimate the degree of openness of the economy, it raises the expectations of foreign exporters, who are then

---

7/ Currently, there are 37 special laws regulating individual commodities. They are: Pharmacist Act, Narcotics Act, Hemp Management Act, Pharmaceuticals Act, Food Hygiene Act, Quarantine Act, Grain Management Act, Fertilizer Management Act, Sericulture Act, Animal Feeds Act, Agricultural Seeds Act, Nursery Trees Management Act, Plant Protection Act, Agricultural Chemicals Act, Livestock Hygiene Act, Foreign Publications Act, Cultural Preservation Act, Film Act, Record Act, Petroleum Industry Act, Energy Conservation Act, State Cigarette Monopoly Act, Ginseng Industry Act, Steamed Ginseng Act, Nuclear Power Act, Regulatory Act on Poisonous Materials, Firearms Act, Electrical Safety Act, Games Conservation Act, Foreign Exchange Control Act, Marine Transportation Act, Fisheries Act, Fishing Vessels Act, Customs Act, Liquefied Petroleum Act and Manufactured Goods Quality Act.

8/ The mechanics are that the "offer sheet" must have approval from the Korean Association of Offer Agents and this often requires consultation with individual industry associations and MTI.

disappointed to find additional administrative barriers, and this disappointment can easily turn into displeasure in trade diplomacy. Second, an overreliance on a single measure, particularly one unweighted with respect to the actual import bundle, can create the impression domestically that the economy is more open than it is in reality. This can be harmful because the public is then more apt to feel that trade diplomacy has been conducted in a one-sided fashion if trade barriers arise abroad for Korean exports. For all of these reasons, greater transparency of the import regime would be helpful.

3.13 In that spirit, Table 3.4 indicates the status of exceptions to liberalization, including emergency and adjustment tariffs discussed later on as well as the tabulation of items subject to selective exceptions i.e., commodities on the diversification and surveillance lists). To this could be added those subject or at least potentially subject to Special Laws. On an items basis, the imposition of selective exceptions has been moderate. As of 1985, approximately 276 items were affected. The number of items subject to either emergency or adjustment tariffs has been exceptionally meager and is testimony to the seriousness of reform.

Impact of Reform

3.14 Although there is a natural presumption that liberalization will result in increased imports of the commodities in question, there are a number of complicating factors, including: (i) the fact that the restriction may not have been a binding one in the first place; (ii) the fact that the level of import demand may change substantially from year to year; and (iii) the fact that some offsetting impediment may arise during the liberalization period. In the Korean case, it is known that newly liberalized goods fall under the same tariff provisions as they did prior to reform (with the careful exceptions of temporary adjustment or emergency tariffs discussed subsequently) and that the weighted average tariff rates for items liberalized over the 1983-86 period will have fallen below their pre-liberalization levels according to data from the Korea Traders Association. Thus, newly liberalized items are in general not subject to additional barriers, and for example, imports of the 305 commodities added to the AA list in July 1983 rose to $1,209 million [9/] a year after reform compared to $911 million the preceeding year. The gain in the second year to $1,442 million was smaller, confirming an earlier finding

9/ The largest increase occurred in intermediate goods ($100 million), of which the most significant gain was registered in electronic parts with most modest gains in machinery; raw materials ($90 million); and consumer durables, garments and other items accounting for the rest.

Table 3.4: EXCEPTIONS TO LIBERALIZATION
(Number of CCCN items at 8-digit level of disaggregation)

	1981	1982	1983	1984	1985
	-- (As of July 1 of the respective years) --				
Tariffs					
Emergency tariff (1)	300	12	69	7	7
Adjustment tariff (2)	-	-	-	14	7
Emergency exceptions (1)+(2)	-	12	69	21	14
-- of which items newly liberalized	-	(5)	(62)	(9)	(4)
Selective Exceptions					
Import diversification list (3)	205	209	174	168	160
Surveillance list (4)	35/a	283	165	142	111/a
-- of which items newly liberalized	-	-	(62)	(23)	(7)

/a Data based on 4-digit CCCN classification.

Source: Ministry of Trade and Industry, Korea Traders Association, and Economic Planning Board.

that imports increase immediately following liberalization but decline in subsequent years.[10/]

3.15 A number of observations can be made with regard to the impact of liberalization. First, there has not been the surge in imports that were to have been expected had the restrictive regime been an hermetically sealed one. Second, public accounts to the contrary there has not been a surge in luxury good importation. Third, there has appeared to be a perceptible increase in the quality of import-competing consumer durables, although not much of a drop in prices. Fourth, the dislocations potentially resulting from new competition appear for the moment to be quite manageable--indeed at present there are only 14 items subject to special tariffs of which only four are newly liberalized. And last, the major beneficiary of liberalization appears to have been Japan, although data is sketchy, and this has added to the already large Korean trade deficit vis-a-vis Japan. The impact of liberalization may be difficult in the 1986-88 period insofar as "back loaded reforms" will increasingly affect commodities covered by monopoly legislation, the producers of which might be expected to be earning economic rents. In July 1986 alone some 39 items, approximately 15% of the total number of monopoly items, will be liberalized.

C. Tariff Reform

Background

3.16 Korea has traditionally relied on quantitative restrictions rather than tariffs to protect import-competing industries, and since export industries have been generally exempt (either directly or via a rebate system), tariffs have fallen mostly on final consumption goods. The weighted average legal tariff rate on all imports is reported to have averaged around 17% in the 1963-67 period and to have risen considerably to 26% in 1967 when quantitative restrictions were eased. Korea thus substituted price for nonprice measures, using among other instruments a Special Tariff System.[11/] The general pattern of tariffs is shown in Table 3.5. The sharp reduction in effective rates in the early 1970s reflected the abolition of the aforementioned special tariffs. Actual tariffs collected, excluding rebates to

---

10/ According to data collected by the Korea Traders Association, the average rate of increase of liberalized imports over the 1978-84 period was 37% in the first year, -4% and -8% for the subsequent two years, respectively, and increases of 12%, 17%, and 26% in years four through six. See Nam (1985) for details on the 1978-83 period. The importation of the 305 items newly liberalized in 1983 grew at a rate of 48% in 1984 (compared to 17% growth for overall imports) and that for the 326 items liberalized in 1984 grew at 31% in the following six months (compared to 9% for total imports).

11/ See Hong (1985) for a description of the supplementary tariffs which were levied on so-called non-essential imports.

Table 3.5: TARIFF TRENDS

	1974	1978	1980	1982	1983	1984
Average tariff rate /a	15.4	16.5	11.3	11.7	13.7	-
Adjusted average tariff rate /b	4.6	8.9	5.6	5.8	7.4	6.4

/a Revenue collections basis for total imports, unadjusted for rebates.

/b Revenue collections basis for total imports, adjusted for rebates.

Source: Nam (1985).

exporters are modest; however, it should be recalled that the denominator of this commonly used ratio is total imports, rather than imports of commodities for domestic sales.

3.17 A more insightful view of the actual impact of tariffs on the economy's import structure can be gleaned from comparisons over time of actual tariffs collected as a proportion for domestic use. Such a comparison is reported in Table 3.6 and it supports the view that effective tariff rates, defined as revenues collected per unit of import for domestic use, have been rising since 1981, as one might expect in a period of trade liberalization, but that overall the rate is modest. There are clearly certain products which benefit from tariff protection, particularly standardized internationally traded products. The agricultural sector is in general more heavily protected than manufacturing, although even within the latter category, there are selected products, such as footwear, where Korean industry, which is highly competitive in external markets, is being provided with a safe domestic market. This issue may well prove troublesome if export competitiveness in these industries begins to lag and the generation of domestic rents becomes imperative for industry survival (see case study on textiles, for example).

Recent Actions

3.18 The second leg of trade reform in the 1980s came to fruition with the Tariff Reform Act of 1984, which included phased general reductions in tariff levels and changes aimed at producing greater uniformity of tariff levels. The simple arithmetic average of tariff rates reached 31.7% in 1982 but declines in statutory tariff levels brought these rates down to 23.7% in 1983 and 21.9% in 1984. The current average tariff of 21.3% falls more heavily on agriculture (28.8% on an items basis) than manufactures (20.3% on a similar basis), and this bias is expected to be maintained. The average is expected to fall to 18.1% in 1988. On a revenue collections basis, 6.4% of total imports in 1984 was taxed, although as a proportion of imports for domestic use collections were 11.0%. In addition to reducing tariffs, the current policy aims to restrict and reduce the industry-specific use of tariff

Table 3.6: ACTUAL TARIFF COMPARISONS
(Tariffs collected as a percentage of imports for domestic use)

Selected products	Effective tariff rates							
	1977	1979	1980	1981	1982	1983	1984	1985
Live animals	16.22	22.58	22.61	22.32	19.83	18.25	17.75	19.49
Vegetable products	4.18	8.30	5.48	5.41	8.06	8.05	9.78	9.10
Animal vegetable fats and oils	16.07	10.79	10.28	10.01	12.32	15.49	15.24	15.14
Prepared foodstuffs	44.29	38.74	23.51	23.72	24.21	27.91	23.15	24.35
Mineral products	0.86	0.81	0.64	0.52	0.69	3.55	4.59	2.57
Chemical products	20.00	18.79	19.33	17.91	20.97	22.58	18.30	18.77
Artificial resins and plastic materials	27.66	22.12	23.37	24.06	24.79	24.40	22.33	21.45
Raw hides and skins leather products	27.20	24.84	17.15	19.51	19.14	22.27	16.51	16.14
Wood and wood products	8.74	3.49	2.96	3.46	3.20	4.05	4.11	4.49
Paper and paper products	12.93	8.43	7.78	8.42	9.84	11.58	9.18	8.90
Textiles	8.71	17.32	11.48	11.65	18.10	15.48	16.77	15.78
Footwear and other manufactured products	22.91	32.59	23.39	47.16	27.79	43.84	31.40	28.35
Stone and cement products	29.81	30.85	27.99	28.11	26.18	28.82	28.41	26.01
Metal products	12.82	12.29	10.21	10.67	12.39	12.55	13.79	13.58
Machinery and electrical equipment	13.10	11.78	8.47	8.20	8.34	9.52	12.95	13.20
Vehicles, aircraft and transportation equipment	12.74	11.49	5.15	3.97	4.10	2.99	7.65	6.55
Optical and audiovisual products	22.24	31.90	15.24	14.96	13.04	14.15	14.86	14.24
Miscellaneous manufactured articles	31.00	27.06	24.36	25.93	30.00	30.72	25.66	23.55
Total Imports	11.21	9.95	7.37	5.92	7.33	8.57	10.19	9.39

Source: Economic Planning Board, Ministry of Finance.

exemptions or rebates as a tool of industrial policy. Rebates were offered to strategic, resource-based, and defense industries prior to the 1984 reforms, and this preference has now shifted to new technology, resource-based and defense industries since then.[12/]

3.19 The objective of policy is to continue to slide the tariff schedule downwards. Beginning with average tariffs on raw materials of 20%, on intermediate (and capital) goods of 25%, and on final goods at 30% prior to reform, the goal is to reach uniform tariffs of 5-10%, 20%, and 20-30%, respectively, for these functional import categories by 1988. If successful, 93.5% of total imports will be subject to scheduled tariffs less than 30% by 1988, compared with only 65.2% in 1980. The progress towards greater tariff uniformity can be seen in Figure 3.2.[13/] By 1984, for instance, the incidence of tariffs above 50% had been reduced from 15.5% of imports to 5.5% and by 1988 it is expected that only 6.5% of total imports will be subject to scheduled tariffs above 30%.

3.20 The overall trade liberalization program promises not only to expose domestic producers to more rigorous market competition but also to reduce the cost of production for export and for domestic consumption. As with any such reform program, Government must be concerned with the short-term adjustment problems facing domestic producers, and for this reason, safeguards have been built in. Among these measures temporarily to assist industries which are exposed to greater external competition are emergency tariffs and adjustment tariffs. The former can be placed on an item to redress a surge in imports. It is reviewed every six months, and takes the form of a tariff surcharge up to 40%. While about 69 items were eligible for emergency tariffs in 1983, that number has been reduced to seven as of July 1, 1985 and only three as of January 1, 1986. Emergency tariffs were used sparingly and carefully to assist domestic producers of newly liberalized items. Items currently eligible for emergency protection are airplanes (total tariff of 30%), casein (30%), and talcum (20%).[14/] Adjustment tariffs are designed to prevent sharp reductions in price and future damage to domestic producers and can be set at

---

12/ Strategic industries included, inter alia, chemicals, cement, basic metals, machine parts, general and electrical machinery, transport machinery, trains, and scientific instruments. Machinery and equipment rebates of tariffs applied to resource-based industries such as primary and power industries. Special industries included shipbuilding (engines) and aircraft (parts) as well as the animal feed and agricultural chemicals industries. New technology industries were designated as machine parts, general machinery, electrical machinery, and electronic materials.

13/ Since no recent effective rates of protection estimates are available, however, it is not possible to compare the progress made in reducing real protection to industry as a result of this round of tariff reform.

14/ Recent deletions from the emergency tariff list include cosmetics, soaps and a range of copper products.

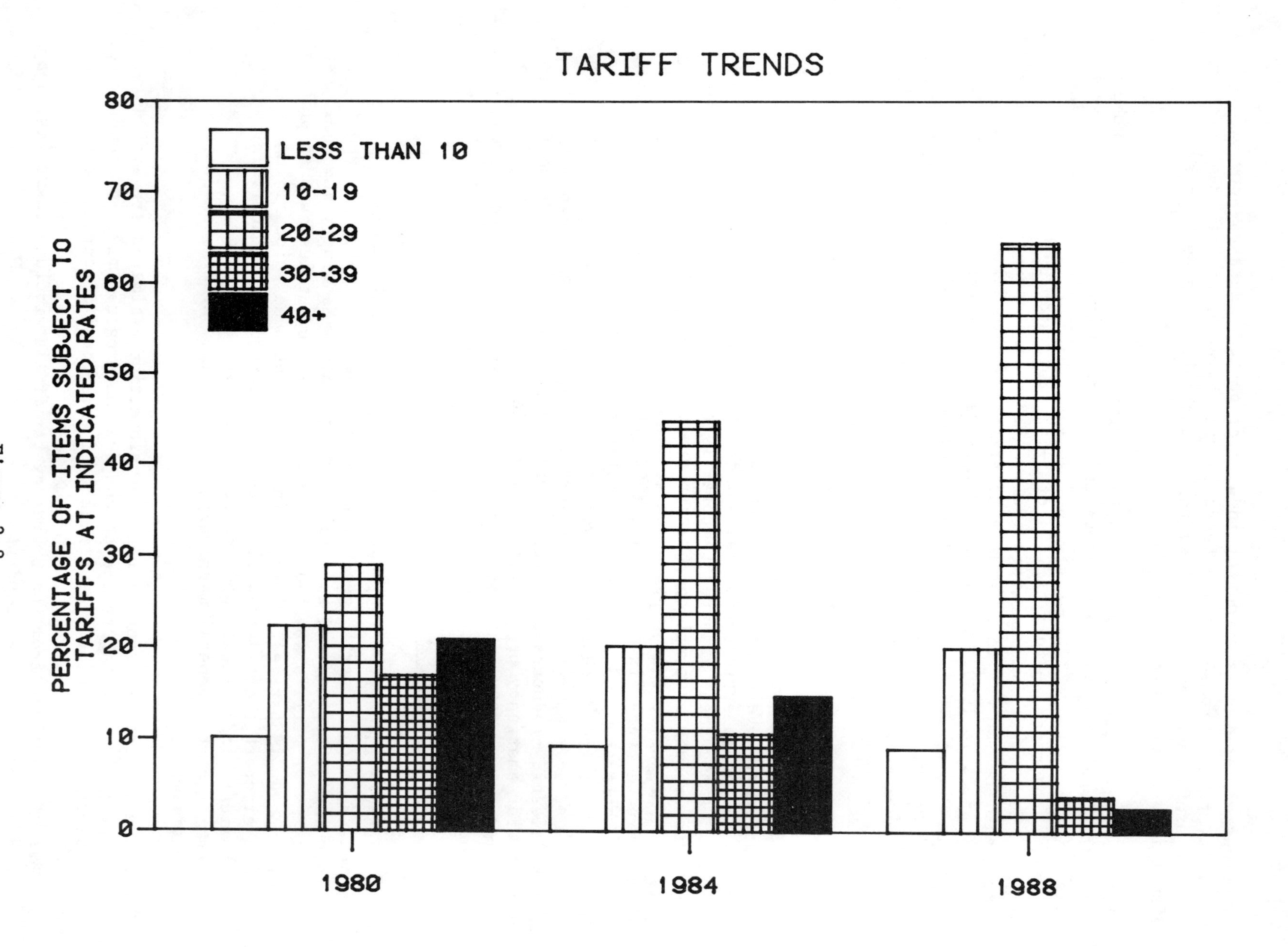
TARIFF TRENDS
LESS THAN 10
10-19
20-29
30-39
40+
PERCENTAGE OF ITEMS SUBJECT TO
TARIFFS AT INDICATED RATES
0
10
20
30
40
50
60
70
80
1980
1984
1988

Figure 3.2

levels of up to 100%. While 14 items were subject to adjustment tariff in 1984, that number had been reduced to a total of 7 items by the second half of 1985 and to only 2 as of January 1, 1986.[15] Table 3.4 indicates number of import items subject to these emergency protection measures. It is noteworthy that these ad hoc interventions have been used sparingly--for example, the total value of imports affected by there actions in 1984 were $15 million--and they have been withdrawn promptly when deemed unnecessary.

D. Overall Assessment

3.21 The important features of the Korean trade liberalization program that have led to its successful implementation have included: (i) its pre-announcement schedule, which has given affected industries advance warning of impending external competition; (ii) the degree to which the schedule has been adhered to, namely, the steadfastness of Government; and (iii) the sparse use of offsetting tariff increases, namely, very limited and temporary applications of adjustment and emergency tariffs and a general trend to both lower tariffs and make them more uniform across industries. Government has made a serious and bold commitment to reform at a rapid pace and has so far weathered the anticipated domestic call for relief.

3.22 Sequencing. The government's decision to liberalize the import regime prior to a more generalized opening of the capital account should be judged against the backdrop of liberalization sequencing episodes elsewhere, especially in Latin America (as discussed more fully in Chapter 3, Volume II of this Report). By allowing producers to adjust to a new import regime without fear of either an appreciating or a highly uncertain exchange rate, both often consequences of capital account decontrol, policymakers can ensure that the export sector expands and that competitive price signals away from import-substitutes are maintained over the liberalization period. By all accounts, maintaining control over the capital account while freeing up the current account is the prudent course--and this has been followed in Korea--because in the absence of very effective sterilization, policymakers may lose some control over the domestic price level, and hence, the beneficial effects of depreciation on export performance may be lost. A step-by-step approach to capital account liberalization is thus advisable.

3.23 Automatic Approvals. While it would, as already noted, be inaccurate to place excessive reliance on the automatic approval ratio per se as an indicator of openness, it should be noted that during the 1986-88 period another 603 commodities will be added to the unrestricted list, of which 53 will be from the list of monopoly items. The recent depreciation of the won, if sustained, should provide additional breathing room for the liberlization program, in particular with respect to the most far-reaching aspects of reform, the addition of new and sensitive items to the AA list, much of which will occur either as of July 1986 or July 1987. Indeed, under certain circumstances, the very favorable exchange rate developments would provide an argument for accelerating reforms; however, given the fast pace already in

---

15/ The items now covered are copper pipes (35%) and electric cables (35%).

effect, the importance of the pre-announcement principle, and the importance of distancing the liberalization program from short-term exchange rate developments, Government would be well advised to stick to its plans. Korea's liberalization program is very much on track and Government merits unequivocal high marks for its effective implementation.

3.24 Special Laws. Since greater transparency of trade policy and a generalized reduction in trade barriers is an explicit goal of government policy, some reform of the extensive coverage of special laws would be desirable. There is a valid case for rules to govern the importation of edibles and health products, for example, but this should not be extended into peripheral products. If, as one suspects, many of the 1800-plus items potentially affected by these laws are not in actuality subjected to any additional restrictions, then it creates a misleading impression on the extent of protection. On the other hand, if constraints are operative, it misleads the public into believing that the economy is more open than it is in actuality. For these reasons, reforms of the special laws should be a high priority.

3.25 Monopoly Items. According to the monopoly and fair trade legislation, there are currently 270 items whose producers are monitored to ensure that they do not use their market-dominating position for unfair gain, namely, that they do not collude in price setting or merger policy to the public detriment.[16/] While such procedures are clearly in the national interest, they raise an interesting issue, namely, why are they needed when many of these product markets could be easily made more competitive through imports. Indeed, a look at the aforementioned 1982 ERP estimates in detail shows fairly widespread levels of substantial protection in many of these commodities.[17/] In a modified application of "second-best policy," Government appears to have protected certain industries and then placed restrictions on the earning of economic rents or, more precisely, increases in those rents when the efficient solution is to open those markets to external competition. While the planned pace of liberalization for monopoly goods exceeds that for overall imports, the continued need for such administrative controls is puzzling for those monopoly goods which are internationally traded.

3.26 Agriculture. Clearly the most protected sector is agriculture, and the estimated increase in ERPs over time are indicative of a sector whose comparative advantage vis-a-vis foreign producers is slipping further. Clearly important economic, political and social tradeoffs are involved in this issue;

16/ Items so classified for purposes of the 1980 Monopoly Regulation and Fair Trade Act include those where one producer accounts for at least 50% of domestic production or 3 producers account for at least 70% and local sales exceed W 30 billion.

17/ Evidence on the degree of effective protection accorded to commodities covered by monopoly legislation can be gleaned by comparing ERP estimates for 1982 (Young and Yoo, 1982) for the 89 (4 digit) commodities involved. Some 28 designated items exhibited negative ERPs, but 26 items had ERPs between 21 and 100% and another 17 items had ERPs above 100%.

however, it is clear that resource misallocation is imposing substantial costs on consumers, indirectly via subsidies to the sector (financed by general taxation) and directly via higher than world market prices for these goods. There are clearly distributional issues at stake between regions, between income groups and between industries.

3.27 Export Industries with Domestic Markets. Various studies report that export incentives, the backbone of Korean trade policy in the 1960s and early 1970s, have been substantially reduced. It is reported that effective subsidy measures which include tax benefits and the subsidy associated with credit preference have fallen from their average level of about 20% in the early 1970s to about half that amount towards decade's end and that other recent studies based on a similar methodology show a further general decline in direct support for exports since 1980.[18/] These trends appear to have been offset, at least partially, by relative increases in protection, as seen in the rise in ERPs for domestic sales of export goods.[19/] New items recently added to the monopoly control list in 1986, for example, include athletic footwear and VCRs, major export items, whose prices apparently will need to be regulated, indicating the charging of monopolistic prices, which is only possible behind some form of protection. In general, it would be advisable for exporters to rely on the exchange rate for major boosts in earnings in the future rather than on domestic markets. This is a particularly sensitive area of trade diplomacy, not only because of protection *per se*, but also because differential pricing between the home and export markets will raise the ire of some of Korea's trading partners.

3.28 Procurement Issues. Although not necessarily a contentious issue at present, the area of international competition for domestically procured items, namely, contracts issued by the public sector, may well be an issue for the future. Often, public procurement is the most susceptible to invisible trade barriers, and it has proven to be a major irritant in US-Japan trade relations. While Korea is a much more open economy than most of its trading partners,[20/] it must continue to pursue liberalization, particularly in products under government control, so as not to be perceived as mercantilistic in its major export markets.

18/ C.H. Nam (1980) reports export subsidies per unit of export value of 20% for 1970 and 10-11% for 1973, 1975 and 1978, while noting of course that all such measures are very sensitive to the interest rate used for comparison to preferential credit. Nam used the difference between preferred export borrowing rate and ordinary bank borrowing rate. See S.S. Lee (1985) for a more recent reassessment.

19/ See Kim and Hong (KDI, 1982) as reported in Hong (1985).

20/ I. Yamazawa and J. Nohara (ed.), *Foreign Trade and Industrial Adjustment in Asia-Pacific Countries* (in Japanese), Institute of Developing Economies: Tokyo, 1985.

E. Future Direction of Trade Liberalization

3.29 The continuation of the present course of trade reforms will not be rippleless, although there can be no doubt that the policies presently in place provide the economy with the best set of incentives to efficient resource allocation. Import liberalization traditionally runs into opposition as it begins to take effect and threatens higher-cost domestic producers. In Korea's case, the eye of the hurricane are the reforms planned for 1986-88 insofar as many previously protected industries, some of them large employers, will be exposed to external competition. The adjustments will be more difficult to absorb if the economy's growth performance lags, which now seem unlikely at least for 1986, or if employment opportunities fail to expand rapidly enough. One might in any event expect the employment issue to loom larger as the export profile becomes more technology-intensive. The answer, however, is not to delay the needed adjustment, but rather to manage it in a way which allows the economy to redeploy resources flexibly. Central to that redeployment is not only labor retraining, but also capital mobility among sectors, a key feature of which is the strength and autonomy of the domestic financial system.

3.30 At the macroeconomic level, if exports lag and self-imposed restraints are kept in place for external borrowing, an argument may emerge for an import slowdown to manage the balance of payments. This would be unfortunate because (i) there are more appropriate policy instruments (i.e., the exchange rate) to affect the balance of payments, (ii) the great merit of phased pre-announced liberalization is its irreversibility, even in difficult circumstances, (iii) the long-term costs to the economy of halting a successful liberalization program will outweigh its short-term attractiveness, and (iv) any _ad hoc_ intervention to manage imports will surely aggravate export market problems in a protectionist world. These reasons notwithstanding, there may well be pressures to act in a protectionist manner, particularly towards trading partners perceived to act unfairly. Protectionism of one sort of another may be coupled with import-substitution plans, especially if the strong import dependence of certain key industries persists and continues to shrink the net foreign exchange earnings of Korean exporters. The danger at present is that Korea, faced with new protectionist barriers to its exports, and still high import-dependence for its exports, will find a selective import-substitution strategy attractive, with its obvious implications for allocative efficiency.

3.31 There is no doubt that the thrust of liberalization should be maintained, with a gradual reduction in embodied exceptions. Korea is moving towards greater transparency in its import policies, a trend that runs counter to and can be imperiled by protectionism in industrialized country markets. It is in this context that worldwide trends are discouraging. In Korea's major markets, for example, various restrictions are affecting approximately 44% of export products accounting for 31% of export earnings in 1984. Additional actions to limit Korea's access to markets will result in economic distortions and dislocations, as efficient product lines will have to be curtailed, and will risk potential backsliding in the opening of Korean markets as well.

3.32 The twin issues of liberalization of domestic markets and fair access to export markets are clearly connected in the context of Korean policy deliberations. The risks, of course, are that hard-won and judicious policies of reform, implemented rapidly and openly, will come undone if Korea is subjected to either harsh protectionism abroad or excessive pressure to open its markets to the limit. Much of the miscommunication in trade diplomacy at present is caused by the changing trade environment in the 1980s; namely, the range of permissible conduct narrowed as trade volumes lagged, global growth rates fell, and the clearly mercantilist policies of some trading nations soured the atmosphere. Nevertheless, since protectionism poses probably the gravest threat to Korean expansion, it must be managed with great care.[21/] Rational and meaningful liberalization will therefore continue to make both economic and political sense.

21/ In the industrialized countries, protectionism is fueled by perceptions of managed trade and unfair pricing. Korean exporters have moved aggressively to exploit any and all market possibilities. Korean zeal to reach export targets has led at times to premature long-run marginal cost pricing or segmented market pricing, and, as a result, to anti-dumping suits brought against producers by the US, Australia, and Canada. While much of the criticism hurled at Korea is a by-product of general trade frustration, Korea, as the latest market entrant in a number of sensitive industries must be acutely aware of the need to manage its trade diplomacy.

CHAPTER 4: STATUS OF FINANCIAL LIBERALIZATION

A. Background

4.01 The financial system in Korea was strongly regulated by the government during the 1970s. Throughout most of the decade, the real interest rates of the banking institutions were negative and a complicated system of preferential loans created a segmentation of financial markets and distortions in the allocation of financial resources. This situation served to retard the development of the financial sector, especially the growth of the highly regulated banking sector, in contrast to the rapid growth of the latter half of the 1960s which had occurred in response to the drastic interest rate policy reform in 1965.[1/] In addition to interventions in the allocation of credit, the monetary authorities exercised extensive supervision and control over the banks' normal operations as well as management, e.g., dividend policy, budget, and personal management. In fact, commercial banks as well as special banks were under almost complete control of the Government, which was in fact the largest shareholder.

4.02 The Government's control over the banks and its intervention in their credit allocation became even tighter in the latter half of the 1970s as part of Government's pursuit of heavy-chemical industry (HCI) promotion. To induce private investment in industries with long gestation periods and uncertain returns, the government had to provide increasingly attractive incentives in terms of preferential loans to a reluctant private sector. The level of bank interest rates were set quite low and severe credit rationing prevailed. The resulting misallocation of resources was reflected in huge over-capacity in the HCI sector and an associated lack of domestic financial resources for other borrowers, such as small and medium industry (SMI). Thus, the need for overall economic liberalization was strongly recognized at the end of the 1970s.

4.03 The Fifth Five-Year Economic Plan (1982-1986), formulated in 1981, noted that active government intervention undermined the efficiency of resource allocation and impeded private initiative, thereby impairing economic flexibility. The Plan emphasized economic liberalization as a means of encouraging greater openness, autonomy, and decentralization of authority in all sectors of the economy. In conjunction with the overall liberalization program, the government started implementing several reforms of the financial system, the ultimate purpose being (i) liberalization of the level and structure of interest rates with the aim of ultimately relying on a market allocation of capital and (ii) the encouragement of competition among financial institutions to increase the efficiency of the financial industry and to enhance the mobilization of domestic financial savings by offering increased real returns to savers.

1/ The high real interest rate policy started in 1965 and ended in 1972 with a return to low real interest rates.

B. Specifics of Reform

4.04 The Korean Government so far has taken a gradual approach to financial liberalization in contrast with the approach which many Latin American countries, such as Chile and Argentina, took in the 1970s (see Chapter 3, Volume II). Government has moved pragmatically to initiate reforms and press ahead until an obstacle is reached, and then has judiciously paused to allow time for designing a revised approach. As a result, although the progress of liberalization has been slow, there have been several significant (and unreversed) reforms during the last five years.

Banking Reform

4.05 As a first step toward financial liberalization, the government started denationalizing commercial banks in 1981 by disinvesting its share. By 1983, the government turned all nationwide city-banks over to private ownership, and reduced its control over day-to-day operations. At the same time, it restricted single shareholders of nationwide commercial banks (except for joint venture banks) to 8% of total ownership.[2/] This restriction was imposed to prevent the banking industry from being controlled by big industrial groups.

4.06 Certain actions were taken to promote the managerial autonomy of banks and to increase their public accountability. Among them were the elimination of Bank of Korea approval for the appointment or dismissal of senior bank officers, greater autonomy in budgeting and in organization. In addition, the authorities abolished in 1982 the system of direct credit controls for each of the Deposit Money Banks (DMBs)[3/] and began conducting monetary policy through reserve requirements, rediscounts and open market operations. Reserve requirements were also significantly reduced and the disparity between those for demand versus time deposits was abolished. While the average required reserves stood at 23% of DMB deposits in November 1979 the ratio was gradually reduced to 3.5% as of November of 1981, and currently it stands at 4.5%. The substantial reduction was intended to give more freedom to banks in their asset management and to ease the strain on bank profits caused by noninterest-bearing reserves.

4.07 To promote competition among banks, two new nationwide commercial banks, joint ventures with foreign banks, were authorized (Shinhan Bank, 1982;

2/ The maximum guarantees and loans for single beneficiary was also limited to 50% of a bank's net worth. Given average city bank capital:asset ratios of 12:1, this implicitly limited lending plus guarantees to about 4% of total assets per borrower. It is not clear whether these limits have been adhered to in the case of lead banks to large industrial groups, for example.

3/ DMBs include the seven nationwide, socalled city banks, as well as local banks, and specialized banks except for the KDB, KEXIM, and KLTCB.

Hanmie (KorAm) Bank, 1983). The restrictions on chartering of non-bank financial institutions (NBFI), in particular short-term finance companies and mutual savings companies, were also relaxed in July of 1985.[4/] At the same time the paid-in capital of the existing city banks as well as special banks was increased to enable them to better compete with other financial institutions, including new entrants. To encourage competition among financial institutions allowed commercial banks and NBFIs to engage in new activities which had previously been prohibited to them, removing artificial barriers to competition between financial institutions through the deregulation of financial services.[5/]

4.08 Another policy measure to increase competition and the overall efficiency of the domestic financial industry allowed foreign banks' local branches to participate in some previously restricted financial services.[6/] Since 1985, foreign banks' branches have had access to rediscount of export bills facilities offered by BOK and as of July 1985 they were allowed to deal in trust business. Starting in 1986, they will be entitled to make use of the rediscount facilities of BOK for all their operations. The government did impose restrictions on foreign banks to allocate 25% of their credit to small-medium firms; however, the requirement is lower than the 35% limit place on domestic banks. The government also has relaxed the entry of foreign bank branches into the domestic market, allowing 53 branches of 46 foreign banks to operate at present. As a consequence, foreign banks now account for about 13.5% of the total amount lent by the banking system, compared with 11.1% in 1982 and 8.9% in 1978. (This share compares to a 3.5% foreign bank market presence in Japan.) Whether this foreign competition has promoted greater efficiency in domestic banking is unclear because of the implicit handicaps to profitability faced by Korean banks, prominent among them being the legacy of

---

4/ Between end-1981 and mid-1985, the number of investment and finance companies increased from 20 to 32, while the number of mutual savings and finance companies increased from 191 to 239.

5/ Commercial banks now issue CDs, sell public debentures under repurchase agreements, and sell the commercial bills they discount. They may now all deal in trust business, issue credit cards and own nonbank financial intermediaries and even securities companies as subsidiaries. This sharpened the competitive edge of commercial banks against NBFIs. Investment and finance companies on the other hand, were allowed to deal with Cash Management Accounts (CMA). At the same time, large securities companies were allowed to participate in some money market activities which had been previously restricted to commercial banks as well as in transactions involving commercial paper carrying a market-determined interest rate, previously the exclusive preserve of investment finance companies prior to 1984.

6/ A second objective of increased foreign banking activity was perhaps additionality in foreign borrowing, although it seems likely that many banks operate under overall country lending limits and that home office and branch office lending operations count equally against those limits.

bad loans which remain from the long history of directed government credit. Foreign banks are reported to have earned 1.3% in net profits (64 billion won) on working capital in 1984 compared to 0.2% for local banks.

Interest Rate Reform

4.09 Perhaps the most important part of the recent financial reform was the rearrangement of the interest rate structure. Although the monetary authorities still maintain the interest rate ceilings on bank deposits and loans, the real rate of interest has been kept positive since 1981. Average real rates during 1981-1984 were about 6% and they currently stand at around 7-10% (see Table 4.1), compared with the negative real rates prevalent throughout most of the 1970s. This is largely due to the marked deceleration of inflation since the latter half of 1981; however, it also reflects government efforts to keep the real rate high by not adjusting nominal interest rates down. In fact, nominal rates have been increased by creating a band in 1984 and raising the upper limit of the band in 1985.7/

Table 4.1: NOMINAL AND REAL INTEREST RATES
(%)

Average for the year	GNP deflator rate of change)	Curb market rate /a		Yields on corporate bonds		Yields on government bonds		Bank lending rate /b	
		Nominal	Real /c	Nominal	Real	Nominal	Real	Nominal	Real
1979	21.2	42.4	21.2	26.7	5.5	25.2	4.0	18.5	-2.7
1980	25.6	45.0	19.4	30.1	4.5	28.8	3.2	22.9	-2.7
1981	15.9	35.3	19.4	24.4	8.5	23.6	7.7	19.2	3.3
1982	7.1	30.6	23.5	17.3	10.2	17.3	10.2	12.0	4.9
1983	3.0	25.8	22.8	14.2	11.2	13.8	10.8	10.0	7.0
1984	3.9	24.7	20.7	14.1	10.2	14.3	10.4	10.5/d	6.5
1985P	3.5	24.0	20.5	14.2	10.7	13.9	10.4	11.5/d	8.0

P Preliminary
/a Bank of Korea survey data.
/b Interest rate on bank loans up to one year.
/c Real interest rate = nominal interest rate minus the rate of change of GNP deflator.
/d Calculated at upper range of band.

Source: Bank of Korea, Economic Statistics Yearbook.

7/ The upper end of the bank was reduced from 13.5% to 13.0% in the fall of 1985 to help boost output.

4.10 Second, the government abolished the complicated preferential lending rate system altogether in June of 1982 and begun to unify the interest rate structure (see Table 4.2),[8] Korea had long ago adopted a preferential lending rate system to subsidize selectively so-called strategic sectors. This system deepened the degree of financial segmentation and, in order to fill the gap, Government had been forced to create new types of preferential loans. In June 1982, the Government unified the bank lending rate to 10% on an annual basis. Third, since 1984, Government has widened the interest rate bands as a first step toward complete liberalization of the interest rate in the future. Commercial banks were gradually permitted to lend at rates with larger margins, presumably depending on a risk assessment of the borrower (see Table 4.3). A more pronounced term structure for deposit rates was also established, as seen in Table 4.4.

4.11 Although the government's stated ultimate goal is the complete liberalization of interest rates, it has approached this goal on a step by step basis. In June 1981 the government established a commercial paper market which is not subject to government control. This market was expected to serve as a bridge between the curb market and the organized financial system and to provide a reference point for setting bank interest rates. The call money rate was also liberalized in November 1984. Korea also opened the domestic security market in a preliminary way by allowing limited foreign investment and provided it with an additional boost in 1983 by deregulating the issue prices of equities.

C. Analysis of Recent Changes in the Financial Market

4.12 The overall size of the Korean financial sector has grown very fast in the 1980-84 liberalization period despite the fact that the national savings rate is generally on par with that recorded in the 1975-79 period, indicating that a substantial amount of savings has shifted from the unorganized market and real property toward financial savings. This is in response to the marked deceleration in inflation and the deregulation of the financial sector which has taken place since 1980. A survey of urban household savings (Table 4.5) shows that the proportion of household savings in financial institutions, especially NBFIs, increased sharply since 1980.

8/ Exceptions are housing loans from public funds and KDB loans financed by foreign funds.

Table 4.2: VARIOUS LENDING INTEREST RATES
(% p.a.)

Source/purpose	1969	1974	1979	1980	1982
Commercial Banks					
Discount or bills up to 1 year	24.5	15.5	18.7	24.5	10.0
Exports	6.0	9.0	9.0	15.0	10.0
Intermediate purchases in foreign currency	6.0	9.0	9.0	15.0	10.0
Equipment for export industries	-	12.0	15.5	21.0	10.0
Equipment for machinery industries	12.0	12.0	15.0	21.0	10.0
Korea Development Bank					
Power, shipbuilding, coal, shipping	7.5	7.5	7.5	7.5	10.0
Industries: Equipment	12.0	12.0	15.0	21.0	10.0
Loans from foreign funds	9.1	9.0	9.1	9.1	7.5
Medium Industry Bank					
Equipment for medium industries (own funds)	20.0	15.5	19.0	24.5	10.0
Equipment for medium industries (government funds)	12.0	10.0	13.5	19.5	10.0
Equipment for medium industries (foreign funds)	8.0	8.0	8.5	8.5	12.7

Sources: Bank of Korea, Economic Statistics Yearbook and Monthly Economic Statistics.

Table 4.3: SELECTED INTEREST RATES ON LOANS
(% p.a.)

End of period	1980	1981	1982	1983	1984	1985
Commercial Banks						
Discounts on bills	19.5	16.5	10.0	10.0	10.0-11.5	10.0-11.5
Loans for exports	12.0	12.0	10.0	10.0	10.0	10.0
Overdrafts	21.5	16.5	10.0	10.0	10.0-11.5	10.0-11.5
Loans on installment savings deposits	17.5	15.5	10.0	10.0-11.5	10.0-11.5	10.0.-11.5
Term loans						
Up to 1 years	19.5-20.0	16.5-17.0	10.0	10.0	10.0-11.5	10.0-11.5
3-8 years	20.5	17.5	10.0	10.0	10.0-11.5	10.0-13.0
8-10 years	21.5	18.5	10.0	10.0	10.0-11.5	10.0-13.0
Other Financial Institutions						
Korea Development Bank						
Equipment for key industries	18.5	16.5-18.5	10.0	10.0	10.0-11.5	10.0-13.0
Using government funds	13.0	13.0	10.0	10.0	10.0	10.0
Using foreign funds	7.5-10.75	7.5-10.75	7.5-10.75	7.5-10.75	7.5-10.75	7.5-10.75
Korea Housing Bank						
Private	10.8	18.5	10.0	10.0	10.0	10.0
Public	8.0	8.0	8.0	8.0	8.0	8.0
Mutual savings and finance companies	-	29.4	20.0	20.0	18.5	18.0
Mutual credit cooperatives	23.5	21.5	13.0	13.0	13.5	14.5
Agricultural cooperatives /a	15.0	15.0	10.0	10.0	10.0	10.0
National Investment Fund (NIF)	18.5-19.5	16.5-17.5	10.0	10.0	10.0-11.5	10.0-11.5
National Housing Fund (NHF) /b	15.0-18.0	15.0-18.0	10.0	10.0	10.0	10.0
Curb market rate	45.0	35.3	30.6	25.8	24.7	-

/a Applies also to loans for fishery.

/b National housing construction.

Sources: Bank of Korea.

Table 4.4: SELECTED RATES ON DEPOSITS
(% p.a.)

	1980	1981	1982	1983	1984	1985
Deposit Money Banks						
Household checking deposits	-	14.4	8.0	8.0	6.0	6.0
Time deposits						
Over 3 months	14.8	14.4	7.6	7.6	6.0	6.0
Over 6 months	16.9	14.6	7.6	7.6	6.0	6.0
Over 1 year	19.5	16.2	8.0	8.0	10.0	10.0
Savings deposits (over 6 month) /a	12.3	14.4	8.0	8.0	6.0	6.0 (12.0)
Installment savings deposits (for households) /a	19.5	16.2	8.0	8.0	10.0	10.0 (13.0)
Certificates of deposit (CDs) /b	-	-	-	-	11.0	12.0
Nonbank Financial Institutions						
Money in trust (2 years)	22.5	20.6	9.7	10.7	10.7	11.9
Investment & finance companies						
Bills resold: 8-29 days	13.3	12.2	8.0	7.5	7.5	7.5
60-90 days	20.4	18.3	9.0	8.5	8.5	8.5
Mutual credit cooperatives						
Time deposits (1 year)	20.7	18.3	9.0	9.0	10.5	11.0
Yield on corporate bonds /c (differential over 1-year time deposits)	30.1	24.4	17.3	14.2	14.1	14.2

/a These are new types of savings introduced in the April 1985.

/b CD-generated loans are 1 percentage point above CD deposit rates.

/c Annual averages for 1980-84.

Table 4.5: URBAN HOUSEHOLD SAVING DEPOSIT PATTERN
(%)

	1980	1981	1982	1983	1984
Organized financial market	29.8	35.8	47.1	55.8	60.6
Banks	22.3 (74.8)	27.0 (75.4)	34.6 (73.5)	38.7 (69.3)	39.3 (64.8)
NBFIs	6.4 (21.5)	8.1 (22.6)	11.6 (24.6)	16.1 (28.9)	20.1 (33.2)
Securities	1.1 (3.7)	0.7 (2.0)	0.9 (1.9)	1.0 (1.8)	1.2 (2.0)
Unorganized financial market	70.2	64.2	52.9	44.2	39.4

Note: Figures in parentheses indicate relative share of savings at each institution as a percentage of total savings in the organized financial market.

Source: Citizens National Bank, "Survey on Households Utilization of Financial Institutions."

4.13 As a result of the deceleration of inflation and Government's recent measures to control real estate speculation, the market for inflation hedges such as land and residential houses has been sluggish. Consequently, the rate of return on investment real estate declined since 1980, although there was a temporary surge in 1983, compared to extremely high returns during 1977-1979. The deceleration of inflation also led to a substantial increase in the real interest rates on financial assets, despite the fact that the nominal interest rates have been adjusted downward since 1980-81. Relative increases in the return on the financial assets compared to that of real assets as well as several financial scandals which took place in the curb market attracted private savings to the financial sector. As a result, the stock of M_3, which includes the liabilities of non-bank financial institutions, has grown at much higher rates than GNP. As seen in Table 4.6, M_3 rose to 60.3% of GNP in 1984 from 45.1% of GNP in 1979. The stock of M_2 has not grown as fast as M_3 though it has grown from 31.6% of GNP in 1979 to 37.7% of GNP in 1984, indicative of the NBFI's faster growth than the banking sector as a whole.

Table 4.6: GROWTH IN THE FINANCIAL SECTOR
(Averages of annual data)

	1970-74	1975-79	1980-83	1984	1985
M_2/GNP	34	32	35	38	39.5
M_3/GNP /a	-	39	51	65	75.9
FA/GNP /b	160	171	274	329	350
National saving/GNP	17.9	25.4	22.7	27.4	28.4
Memo Items (av. growth rate)					
Real growth of M_3 (%) /c	-	12.2	12.8	18.1	23.0
Real growth of DC (%) /d	-	10.9	17.5	15.3	19.0

/a M_3 is defined as the sum of M_2 plus deposits at non-bank financial institutions and commercial bills sold and certificates of deposits and debentures issued by deposit money banks.

/b FA refers to total financial assets of the economy which include assets of financial institutions but excluded foreign assets and trade credit.

/c Deflated by the GNP deflator. Average of annual rates.

/d DC refers to domestic credit extended by the Bank of Korea, deposit money banks and non-bank financial institutions. Average of annual rates

Source: Bank of Korea, Economic Statistics Yearbooks.

Structural Changes

4.14 Two distinctive structural changes have taken place in the Korean financial sector. First, the share of commercial banks in deposit taking has been continuously shrinking with the rapid expansion of NBFIs. Second, securities markets, such as for commercial paper and corporate debentures, have grown very rapidly. There are several reasons for the relative decline of the banking sector and expansion of NBFIs. When the Government in the early 1970s encouraged the establishment of the non-banking sector with the purpose of moving funds from the curb market to the organized financial market, NBFIs were accorded preferential treatment, viz., they could offer higher interest rates to their depositors than could banks. Since NBFI mobilized funds by issuing deposit-type liabilities from large savers and extend credit mostly to large borrowers, their intermediation costs may be lower than for banks which must mobilize funds from a number of small depositors. This undoubtedly aided their profitability and growth.[9/]

9/ It is widely reported that large industrial firms were major participants in the NBFI market, as borrowers, lenders, and owners of NBFIs.

4.15 NBFIs. More important, however, is the fact that NBFIs have been much less regulated with respect to their asset management and lending rates than the tightly controlled banks. In pursuing financial liberalization, the government has taken a gradual approach in which it deregulated first the interest rates at the NBFIs and then relaxed further the control over the asset management of these institutions.[10] At the same time, the authorities eased restrictions on entry into the nonbanking sector and as a result, the number of NBFIs increased very fast and the share of total deposit liabilities held by NBFIs increased from 26.7% in 1980 to 42.4% at the end of 1984 (see Table 4.7). In terms of monetary sources of borrowing by the corporate sector, the expansion of the NBFI role is even more impressive, as its share increased from 37% in 1977-79 to 64.4% in 1984, while that of banks decreased from 63% to 36% in the same period (see Table 4.8).

4.16 Recently, therefore, the Government, concerned about the continuously shrinking asset share of the banking introduced more attractive deposit instruments in the banking sector. The Government has been particularly concerned about the poor profitability and uncompetitiveness of commercial banks, buffeted by a declining deposit share while at the same time burdened by nonperforming loans, because it recognizes that the poor performance and profitability of banks places a significant impediment on further liberalization of the banking sector. Recent actions aimed at shoring up banks' profitability include the introduction of new types of savings deposits, which pay higher interest rates than NBFI instruments, and special low interest (3%) central bank loans. Since the former policy change in April 1985, savings in the banking sector have increased quite rapidly, reversing a steady decline of the past decade, seen in Table 4.7.

New Instruments and Markets

4.17 New Instruments. The markets for corporate bonds and commercial paper have also grown rapidly since 1980. Compared to the growth of the stock market, these markets showed a remarkable surge during the last five years. This growth seems to be largely due to two factors: first, as a consequence of the Government's tight monetary policy, the corporate sector could not obtain sufficient financing from monetary institutions, forcing firms to rely either on the direct credit market or nonmonetary institutions. Second, the

---

10/ Although there is no formal interest rate ceiling in NBFIs, the interest rates are still set following the guidance of the Ministry of Finance.

Table 4.7: DEPOSIT SHARE OF BANKS AND NBFIs
(Billion won) /a

	1974	1978	1980	1982	1983	1984	1985 Jun
Bank /b	21,933 (85.7)	68,831 (79.3)	115,375 (73.3)	188,474 (64.6)	220,956 (60.9)	248,188 (57.6)	275,127 (58.6)
Non-bank	3,666 (14.3)	18,014 (20.7)	42,085 (26.7)	103,305 (35.4)	144,745 (39.1)	182,679 (42.4)	194,714 (41,4)
Investment and finance companies	1,622	9,407	20,984	42,273	54,971	70,118	64,990
Investment and trust companies	53	2,413	6,351	27,683	36,536	43,129	49,027
Mutual savings and finance companies	507	1,607	4,000	9,566	14,743	19,917	23,878
Life insurance companies	978	3,514	9,427	22,087	33,634	47,383	54,368
Others	506	1,073	1,323	1,696	1,861	2,286	2,451
Total	25,599	86,845	157,460	291,779	362,701	430,847	469,841

/a Figures in parentheses are percentage shares.

/b Includes money trust, commercial bills and demand deposits.

Source: Ministry of Finance, "Fiscal and Financial Statistics."

corporate bond and commercial paper markets were not subjected to government regulation--interest rates are determined relatively freely--and consequently, the yields in these markets have been higher than those offered by financial institutions and thus have been more attractive to investors. In 1984, for example, corporate yields aveaged 14.2% compared with 11.4% for trust certificates or 7.3% for time deposits. The rapid growth of this market was also facilitated by the banks' guarantees of corporate debenture. As seen in Table 4.9, practically all new corporate bond issues have been guaranteed by financial institutions, mostly by banks. A general observation on the Korean financial system is that the success of new instruments, such as bonds, and of new institutions, such as NBFIs, is implicitly to some extent on the safety net provided by the formal banking sector, with vestiges of public support. The issue of moral hazard still abounds in the Korean context. Recognizing the danger of this in a highly leveraged corporate environment, Government has tried to develop the equity market.

Table 4.8: SOURCES OF FUNDS BY CORPORATE SECTOR /a
(%)

	1966-71	1972-76	1977-79	1980-83	1984
Borrowing from monetary sources	40.5	43.3	50.4	33.3	54.4
Banks	(31.9)	(29.2)	(32.0)	(17.4)	(19.4)
Non-banks	(8.6)	(14.1)	(18.4)	(15.9)	(35.0)
Securities	14.2	18.5	22.5	25.5	27.7
Bonds	(0.7)	(1.9)	(6.5)	(10.3)	(12.5)
Stocks	(12.0)	(15.2)	(14.5)	(7.7)	(12.1)
Capital paid in	(1.5)	(1.4)	(1.5)	(7.6)	(3.1)
Corporate bills	-	1.9	2.4	5.6	-0.7
Government loans	0.7	0.0	0.1	2.6	0.7
Borrowings from abroad	30.9	26.3	13.0	15.7	22.4
Total	100.0	100.0	100.0	100.0	100.0

/a Includes noncorporate enterprises and government enterprises since 1980.

Source: The Bank of Korea, Flow of Funds Accounts and Economic Statistics Yearbook.

Equity Market

4.18 The stock market was sluggish between 1980 and 1985, despite government efforts--including more tax incentives and issue price deregulation--to the contrary. The reasons for this lethargy include: (i) poor corporate performance during the economic recession following the second oil shock; (ii) the volatility of stock prices; and (iii) the fact that other financial assets (i.e., debt) have been more attractive with deregulation of the financial institutions. Structurally speaking, Korea's stock market is rather thin, accounting for less than 8% of the assets of DMBs. Much of Korean industry is owner-managed, and despite more vigorous reporting requirements (and the recent creation of rating services), the stock market is still embryonic. As a result of the economic recovery and a series of measures designed to foster the securities market, there has been a slight recovery in stock issuances since 1983. However, the number of shareholders is still below the level of 1978 and the number of listed companies has been stagnant since then as well.

4.19 International liberalization of the equity market is also modestly underway. Since 1981, five foreign investment trusts have been established and in addition, the Korea Fund, a closed-end mutual fund was established in 1984 and was listed on the New York Stock Exchange. These efforts were aimed at increasing the efficiency of the securities market operation and boosting the supply of equity capital in these markets.[11/] An additional means of promoting the internationalization of Korea's capital (particularly equity) market has been Government's authorization for large firms to issue convertible bonds and depository receipts in international capital markets.

Table 4.9: OFFERINGS OF CORPORATE BONDS BY TYPE
(Billion won, %) /a

	1972-76	1977-79	1980-83	1984
Total	32.2	375.8	1,384.6	1,804.1
Guaranteed	25.9 (80.4)	349.6 (93.0)	1,360.8 (98.3)	1,664.7 /b (92.3)
Not guaranteed	6.3 (19.6)	26.0 (6.9)	22.2 (1.6)	139.4 (7.7)

/a Figures in parentheses represent percentage shares.
/b Includes mortgage bonds.

Source: Securities Supervisory Board, Monthly Review, various issues.

D. Credit Allocation Policy

4.20 As already mentioned, the Government fully recognizes the importance of the market determined allocation of resources to increase allocative efficiency and has announced its intention to reduce its intervention in the banks' credit allocation. In practice, there has been some effort to reduce the types and magnitude of policy directed loans. In 1981, for example, the government abolished some categories of preferential lending to allow banks greater discretion in their lending operation. The size of the National Investment Fund, which is one of the major vehicles for directed credits, was also reduced. However, despite the Government's philosophical redirection, some degree of intervention in the bank's credit allocation is inevitable. Recently, intervention has taken the form of credit reservation for the SMI sector, an "affirmative action" kind of policy to offset past preferences to large firms, and a prudent measure in light of the grip of conglomerates over the banking sector by virtue of both their asset concentration as well as

---

11/ Korea International Trust, 1981; Korea Trust, 1981; Korea Growth Trust, 1985; Seoul International Trust, 1985; Seoul Trust, 1985. The total size of these trusts is $140 million, and taken together with the Korea Fund, have brought about $200 million into the Korean equity market. While small by international standards, this accounts for about 5% of the value of traded stocks.

their ownership position (see Table 4.10). Indeed, insofar as capital markets are immature, and still characterized by credit rationing of some kind and market segmentation, a prudent course of liberalization is advisable as long as there is "light at the end of the tunnel." Government appears to be taking this kind of measured approach.

Table 4.10: CONGLOMERATE OWNERSHIP OF BANKS
(%)

	Cho Heung	Korea First	Hanil	Bank of Seoul	Comm'l Bank
1. Hyundai	2.14	9.35	7.27	11.93/a	-
5. Daewoo	1.23	23.82/b	2.22	5.29	4.48
3. Samsung	8.34	5.69	9.72	-	15.97/b
4. Lucky Goldstar	1.71	5.30/a	5.87/a	-	-
5. Hanjin	-	-	8.45/b	-	-
6. Taekwang	3.77	-	-	4.56	-
7. Ssangyong	5.57	-	-	-	-
8. Daelim	-	-	9.29/b	-	-
9. Shindongah	7.98	7.24	-	9.90	-
10. Dong Ah	-	-	10.03	-	-
11. Hanil-Kukje	4.05	2.18	3.69/b	-	1.91
Memo Item: Top 10 Ownership	39.79	54.58	56.54	31.68	22.36

/a Quasi-lead bank.
/b Lead bank.

Source: Business Korea

New Guidelines

4.21 As a first step in the liberalization process, Government opted to correct the bias in bank loans concentration which had resulted from past public sector intervention. For example, heavy and chemical industries and large firms were the major recipients of government-induced investment allocations in the 1970s. However, the effort to correct the undesirable consequences of past policy called for yet another restrictive policy in credit allocation; namely, the government has pushed banks to allocate more credit to small and medium firms since 1980. It mandated that 55% of the increase of any local banks' credit and 35% of the nationwide city banks' credit should go to small-medium firms.[12/] In addition, the government in 1984 froze the share of the 30 largest conglomerates in total bank credit at the level of end-1983. Moreover, firms whose debt-equity ratios exceeded 500% (mostly large firms) were not allowed further access to bank credit. As a result, the availability of bank credit to small and medium firms has increased since 1980, and correspondingly, the average cost of borrowing of small and medium firms also has decreased.[13/]

4.22 By mid-1985, however, this policy line had to be eased for several reasons. First, the pursuit of this policy together with tight monetary policy put double pressure on the large corporate sector in terms of reduced access to credit, a good deal of which is needed on a rollover basis by highly leveraged firms. Second, the restructuring of financially troubled industries (i.e., shipping and overseas construction) required greater credit access for firms involved, mostly large firms. Third, the recent poor performance of exports put pressure on authorities to relax credit restrictions on large firms in order to encourage investment and promote sagging exports. By May 1985 most restrictions on bank credit to large firms had been rescinded and the share of credit going to large firms rebounded once again (see Table 4.11).

Financial Restructuring

4.23 Much of the investment in HCI which was carried out under government direction and inducement during the late 1970s resulted in idle capacities. Government thus felt obliged to assume an important role in the ensuing industrial restructuring efforts, by facilitating mergers, scrapping unprofitable plants, and realigning industries. This forced the Government to intervene heavily in recent years, irrespective of its announced intentions to the contrary, in financial allocation decisions.[14/] In addition to this restructuring of HCI, government has bailed out several declining industries to reduce the immediate impact of bankruptcies on employment and growth as well as the banking system *per se*. The banking sector, recently denationalized de jure, is in reality not yet able to cope with "financial workouts" on its own, and in fairness, perhaps should not be expected to bear the cost by itself of past interventionist policies. Since the slate cannot easily be wiped clean, the challenge to Government is to establish a process by which banks gradually take increased responsibilities for asset management, even in difficult

---

12/ Small-medium firms are defined as those whose total assets are less than Won 1 billion, the total number of employees is less than 300 in the case of a manufacturing industry and 200 in the case of a service industry.

13/ See Chapter 5, Volume II.

14/ Major examples of these interventions were rationalization programs for the shipping and overseas construction industries, earlier reorganizations in the machinery industry, and government involvement in petrochemicals and fertilizers. See Chapters 2 and 5 for details.

Table 4.11: SHARE OF SMALL FIRMS BORROWING IN THE NET INCREASE OF DMB'S CREDIT
(Unit: 100 million won)

Year	Total credit of DMBs (A)	Credit to small and medium firms (B)	SMI share (B/A)
1970	1,594	457	28.7
1971	1,971	495	25.1
1972	2,785	900	32.3
1973	3,895	1,502	38.6
1974	8,403	919	10.9
1975	4,777	1,632	34.2
1976	8,194	2,227	27.2
1977	9,841	3,611	36.7
1978	19,200	5,112	26.6
1979	23,488	8,512	36.2
1980	32,266	10,058	31.2
1981	37.506	14,283	38.1
1982	42,708	12,724	29.8
1983	39,245	15,440	39.3
1984	38,286	15,251	39.2
1985	58,463	19,641	33.6

Source: Bank of Korea.

circumstances, and begin to allocate capital purely on the basis of profitability.[15/]

4.24 As the above factors indicate, there have been significant constraints facing Government in its expressed desire to reduce its intervention in the credit allocation process of banks. Under current adverse circumstances, cutting the denationalized banks loose may well be counterproductive. In this regard, it should be noted, as seen in Table 4.10, that major conglomerates tend to hold controlling interests in a number of city banks and that without continued close supervision one could easily see the integrity of the financial sector being compromised, as has been the case in other countries pursuing financial reforms. This risk is higher when distress financing becomes prevalent, as is the case when interest rate decontrol proceeds. Nevertheless, continued progress in establishing bank autonomy vis-a-vis asset management is an important prerequisite for continued domestic financial reform, and ultimately, for capital account decontrol.

E. The Progress of Financial Reform

4.25 The growth of Korea's financial sector during the last five years has been quite impressive, and is primarily due to high and stable real interest rates on financial assets as well as the deregulation of financial institutions, which has, in turn, encouraged competition and the creation of new types of financial instruments. The Korean financial system has become more competitive and efficient because of deregulation. Greater competition among financial institutions, development of various types of financial assets, including the development of securities markets, provided more diversified sources of corporate finance as well as household savings. This contributed to a reduction in both segmentation of financial markets and distortions in financial allocation (see Chapter 5, Volume II of the Report). This high growth of the financial sector is seen not only in terms of size, but also in terms of efficiency, and seems to have been based, to a large extent, on the rapid expansion of the less regulated non-banking sector, and, probably to lesser extent, on the liberalization of the banking sector per se.

4.26 The most significant developments on the banking side are the abolition of the preferential rate system and the maintenance of high real interest rates with increasingly large margins allowed depending on maturity and risk of loans. While this has served to reduce the distortion in financial costs among different sectors, rates are still not freely determined nor are they a properly risk-adjusted.[16/] In terms of the autonomy of bank credit

---

15/ It may well be argued that erasing the egregious errors of the past on bank balance sheets, although costly on the public treasury, may be beneficial in reanchoring the banking system.

16/ Major firms, for example, may be able to borrow at the lower end of the interest rate band if they hold compensating balances, and the rates are based in large measure on corporate identities, not the ultimate uses of the funds.

Table 4.12: PROFITABILITY OF FINANCIAL INSTITUTIONS
(Profit ratios) /a

	1977-79	1980-82	1983	1984
Nationwide city banks	0.95	0.55	0.13	0.26
Local banks	1.26	0.83	0.25	0.74
Investment & finance companies	4.47	4.13	1.50	-
Merchant banking companies	3.93	5.03	2.70	-
Investment & trust companies	0.88	8.83	9.40	-

/a Net profits divided by total asset.

Source: Federation of Korean Banks, The Analysis of the Efficiency of Banking Industry and Suggestions for Higher Efficiency (Korean), 1985.

allocation procedures, moreover, there has been labored progress. The effort to liberalize the credit allocation process has been limited by the acute restructuring demands of some key industries and the accumulated, substantial amount of nonperforming loans. Thus, the situation of commercial banks will require government policy initiatives to relieve banks of the burdens of the past and allow them to be responsible for their future lending decisions. As has been shown, the profitabiliy of commercial banks, especially the top nationwide city banks, is much poorer than the profitability of nonbanking institutions, which reflects the relatively freer asset management and interest rate policies of the NBFIs (see Table 4.12). The banks' profitability would have been even lower had the Bank of Korea not paid interest on commercial bank reserve requirements. The problems of nonperforming asset are substantial impediments to further liberalization of the banking sector, which is in and of itself a desirable goal, but moreover is an important precondition for dealing with the problem of industrial restructuring and also is a prerequisite for fuller financial liberalization.

F. Future Direction of Financial Liberalization

4.27 The Korean Government continues to be confronted by several problems that require resolution in order for further liberalization to proceed. The first concerns the nonperforming assets of nationwide commercial banks. The government so far seems to have approached the industrial restructuring of declining industries by shifting the burden of adjustment to commercial banks, avoiding the short-term consequences of bankruptcies, and the possible losses in terms of employment and growth. It is now recognized, however, that the accumulation of nonperforming loans limits the autonomy and the efficiency of the banking industry and raises the future adjustment costs to the economy. The current approach of Government seems to be a blend of fiscal and financial solutions, i.e., the provision of tax exemptions to facilitate the disposal of real estate holdings of firms which are held as collateral by the banks, and second, increased low-cost rediscount loans offered by BOK to commercial banks to reduce the burden of nonperforming loans. The fiscal impact of the former intervention serves to shift some of the adjustment cost to the taxpayer at large, a result generally inferior to allowing shareholders to absorb losses. The second approach runs the perhaps larger risk of breaching monetary targets, and rekindling inflation. Therefore, a more satisfactory, long-run solution to the problem of bank profitability must be found. Without such a solution, continued government involvement in successive industrial restructurings and even large firm bailouts appears to be inevitable, jeopardizing one of Korea's major policy objectives--the eventual removal of government from industrial decision making.

4.28 The second problem the government faces is to provide a proper supervision and regulatory framework for NBFIs in order to secure the stability of these institutions and the entire financial system. While NBFIs have been unconstrained as to management of their assets, (i.e, their lending to particular classes of firms or industries), they have implicitly benefitted from the Government's lingering involvement in the formal banking sector. There is a widespread belief that firms will not be allowed to default to banks, and that this implicitly protects the NBFIs' assets, in yet another twist of the moral hazard principle. This encouraged the explosive expansion of this sector of the financial system. However, as the NBFI share and importance in the overall financial system increases, it becomes important to provide proper financial supervision and coordination, especially considering the coexistence of the NBFIs with highly regulated banking sector. Given the expectation that a firm will be bailed out by banks if it gets in trouble, NBFIs can engage in distress financing, which not only will retard industrial restructuring but will also aggravate the overall stability of financial institutions. A supervisory framework might help ameliorate this problem; however, it is important that this framework be structured in such a way that it does not replicate strong intervention as has been the case in the banking sector.

4.29 A third critical issue that policymakers face in pursuit of further financial liberalization is the high debt-equity ratio of the corporate sector, and relatedly, slow development of equity markets. Although high real interest rates, tight credit control and some government jawboning has served to reduce indebtedness in the large corporate sector, it is still very high. The highly leveraged corporate sector is very vulnerable to economic shocks and this may prompt frequent government interventions in the banking system in order to avoid financial disruptions and economic dislocations. Moreover, as Korean industry seeks to adjust to export market developments and tries to rapidly redeploy resources into new products, the overindebtedness of its industry will exert a drag on the economy. Therefore, it will be important to develop a well-functioning equity market as one means of reducing corporate indebtedness.

4.30 The development of the equity market will also be important to substitute private risk capital for the implicit role played by government over the years as risk partner of industry. As this role diminishes, with the shrinking share of government-controlled banks in corporate finance, further liberalization of financial sector, and industrial policy reforms, stronger capitalization of industry is essential. Although the securities market has grown markedly recently, its growth is mostly due to the expansion of the (guaranteed) bond market. The stock market, already small in size, has been quite sluggish during the last several years and greater efforts to encourage its growth may be necessary. An encouraging development on the equity side are new venture capital trusts, essentially domestic mutual fund variants for direct investment in new businesses, particularly small, technology-intensive firms. However, as long as debt capital is easily accessible to the large corporates regardless of the underlying strength of the company or the ultimate use of funds, securing more equity capital will appear relatively unattractive to them.

4.31 With respect to all these developments, the Korean government may have to think about the structure of the financial sector as Government's role in financial matters begins to recede. The Government has already relaxed the strict demarcation between NBFI and commercial bank business domains to encourage competition by allowing both to cross previously exclusive business lines. At the same time, many specialized banks now operate in virtually the same businesses as do commercial banks. But the creation of new institutions and instruments does not obviate the need to deal with the problems of the existing banking sector, nor does it reduce the imperative of continuing to reduce market segmentation and credit rationing as well as the inefficiencies of artificial interest rate differentials. All this is to suggest that a clear direction on the role of participants in the financial system needs to be developed. An important corollary is that these decisions should precede external financial liberalization, as discussed in Chapter 3, Volume II in some detail.

4.32 An important corollary is that these decisions should precede external financial liberalization because, as noted in Chapter 3, Volume II, domestic reform should be substantially completed before opening the economy to unimpeded international flows. Positive real rates do exist, but well functioning capital markets ready to compete with foreign markets will require less segmentation as well as a wider range of maturities and instruments (e.g., to hedge exchange rate risk) than now available. Moreover, for the efficient conduct of monetary policy, a strong ability to sterilize excess capital inflows would be needed, perhaps in the form of open-market instruments.

CHAPTER 5: CURRENT INDUSTRIAL POLICY ISSUES

A. The Economics of Industrial Policy

Conceptual Framework

5.01 Although widely used, the term "industrial policy" has no standard definition. In the industrialized countries, and particularly in the United States, it is usually refers to highly interventionist policies--attributed to, if not necessarily practiced by, countries such as Japan and France--that accelerate the growth of some sectors and manage the decline of others. The notion of "picking winners and identifying losers" is associated with this view of industrial policy, which because of the potential distortions implied by government control over the structure of output, is widely considered to be inefficient relative to the market mechanism.[1/]

5.02 In the development context, industrial policy does not necessarily connote industrial targeting and is understood to subsume a wider array of policies that affect production and trade. Indeed, industrial policy encompasses all policies that promote industrialization, and includes, inter alia, measures that affect import substitution and export promotion, exchange rates, trade barriers, industrial financial and fiscal policies. In this chapter industrial policy is used in its broadest sense--as a generic term for the spectrum of policies that may be relevant to furthering Korea's industrial development. For taxonomic purposes, policies that affect industry (i.e., the incentive structure of an economy) can be grouped under three main headings, defined as the incentive regime, functional incentives, and selective incentives.

5.03 Incentive Regime. The most important aspect of industrial policy is the general pattern of production incentives formed by fiscal, financial, and trade policies. The broad structure of these incentives determines, for example, whether a country's development strategy is import substituting, neutral, or export promoting. In smaller, open economies, the structure of trade policies is often the primary determinant of the pace and character of industrial development. In Korea's case, the outward-looking orientation of the incentive regime has played a central role in the country's industrial takeoff, and is widely cited as a model regime.[2/] The success of Korea's incentive regime has often been abscribed to its "modestly pro-export" bias. As described in Chapter 4, Korean incentives were far from neutral, especially in the early phases of the big takeoff but export promotion policies were generally more than sufficient to offset other biases against export activities.[3/] More recently, the Korean incentive structure has begun to move towards greater neutrality.

1/ For example, see Krugman (1984), or Shultze (1983).

2/ See, for example, Krueger (1979), Bhagwati (1979), Balassa (1985).

3/ Westphal (1985).

5.04 Functional Incentives. Beyond establishing a general development climate, public intervention may be warranted in markets for specific factors and to preserve competitive conditions in output markets. From an analytical viewpoint, market interventions may be justified by the absence of, or imperfection in, markets for "commodities" such as knowledge, technology, human capital, and certain types of physical infrastructure. To increase the supply of these critical inputs, governments may invest in infrastructure, manpower development and research, and provide incentives for private research and venture capital. Policies in these areas are becoming more important in Korea, but care must be exercised to see that public interventions are limited to cases of real market failure. The Government has moved to indirectly strengthen these private markets by taking steps to spur competition and to develop new financial institutions.

5.05 The common feature of functional interventions is that they are not designed to affect the sectoral structure of the economy, although they may have an uneven sectoral impact in any particular application. Their purpose, rather, is to offset some systemmatic market failure or pattern of distortion in particular markets. Clearly, at any particular stage of the development process some markets are more prone to imperfection than others. The challenge to Korean policy-makers is to develop the institutional mechanisms that (i) identify imperfections and accurately assess their extent; (ii) provide effective remedial action of the appropriate nature, strength, and duration; (iii) ensure that the negative "side-effects" of the intervention do not outweigh its positive benefits; and (iv) phase out remedial policies as the affected market matures. Timely phase-out, in particular, has proven exceptionally difficult in many countries as market agents have become chronically dependent on subsidies.[4/]

5.06 Selective Incentives. Frequently, public interventions also involve specific sectoral, sub-sectoral or industry promotion objectives.[5/] Government will undertake to identify "sunrise" or "strategic" industries and support these with policies ranging from direct investment to subsidies, technological support, and protection. They will also intervene to manage the withdrawal of factors from "sunset" industries with policies ranging from nationalization to adjustment grants, loans and guarantees, and the orchestration of mergers and cartels. Intervention in favor of emerging industries is a perennial developing country issue, and indeed sector-specific promotion programs played an important role in Korean industrial policy during the 1970s. As newly industrializing countries like Korea mature, they will increasingly face industry-specific pressures to intervene in support of declining industries, mirroring what has become commonplace in recent years in industrialized countries. Some pro-interventionist approaches aimed at

---

4/ In the industrialized countries, this is seen most vividly in agriculture, where farm subsidy programs, below-cost access to water and credit, and import protection have proven difficult to dismantle.

5/ In addition to policies which aim to achieve industrial objectives, policies pursued to reach other objectives (for example, employment creation or distributional equity) also have ramifications for industry.

preventing job losses and protecting international market shares are based on the mistaken view that there is no such thing as a declining industry, only antiquated technology.

5.07 Dynamic Considerations. It is often argued that market-led development is efficient in a "static" sense, but that government leadership is needed to establish industries that are "dynamically" appropriate. This is the argument predominantly used to justify selective intervention in favor of strategic industries. It is true that government policies, particularly as they affect physical and technological infrastructure, have to be sensitive to dynamic considerations, especially for countries aiming at rapid change. It may also be true that governments have an advantage in earlier stages of development in gathering information needed for long-term decision-making. Indeed, in Korea the government substituted for what it considered lack of the long term entrepreneural view with export guidance and encouraged the formation of large, integrated economic agents with whom it could deal efficiently.

5.08 In the 1970s, government also become a risk partner of industry in areas where it felt that dynamic advantage would be developed, but not without direct intervention; however, this policy turned into excessive, prolonged support for some unprofitable activities in the late 1970s. But in most economies, especially rapidly growing economies such as Korea, entrepreneurs and markets become keenly concerned with dynamic factors, and this implies that governments should do their best to progressively confine their interventions to providing infrastructure and disseminating information, while leaving ultimate production and investment decisions to market agents. Barring externalities or market imperfections, the fact that an economy is progressing dynamically does not itself constitute a reason for substituting government planning for market decisions.[6/]

5.09 The theoretical advantages of neutrality notwithstanding, a government's practical choices are constrained by the capabilities of existing economic institutions, which in turn depend in part on the history of past intervention. For example, the government's role in allocating credit in Korea over the past quarter century has retarded the development of financial markets and has as a result mandated continued intervention in certain areas. These constraints act as a brake on the adoption of more neutral policies, but they do not obviate the need for a shift in that direction. Indeed, more neutral policies can, if properly designed and implemented, have the added

---

6/ As a case in point, the existence of "learning curves"--which suggest that productivity increases with production experience--does not, in itself, necessarily justify government aid. Private markets can and do undertake activities in which initial losses are accepted in anticipation of future profits. Therefore, government support is theoretically necessary only in the presence of some additional complicating factor, such as imperfection in capital markets, or evidence that the benefits of learning cannot be fully appropriated by the firm that undertakes the initial investment. In practice, however, this strict test is not applied, and governments often simply provide protection for an infant industry until its costs of production fall to internationally competitive levels.

benefit of promoting the development of institutions that can improve the economy's efficiency in responding to market signals and allocating resources.

5.10 While many conventional arguments for government intervention cannot be rigorously defended, there are also exceptions to the rule that market decisions are optimal. These exceptions center on market failures, externalities, and strategic behavior, and can be fashioned into rigorous arguments for subsidizing factors and products, supporting infant industries, and restructuring declining industries.[7] At the same time, there is considerable evidence that: (i) these exceptions are less pervasive than policy-makers often believe; (ii) compensating interventions to offset initial market distortions may exacerbate rather than ameliorate efficient resource allocation; and (iii) interventions are often not formulated carefully enough to avoid secondary distortions in economic decision-making. An illustrative list of industrial policy objectives, possible underlying market imperfections, and common policy remedies is presented in Table 5.1. Some of the issues covered by this list are reviewed in greater detail in the Appendix to this chapter which addresses the theory of industrial policy in the Korean context.

The Korean Setting

5.11 The formulation and conduct of industrial policy in Korea was based on a favorable overall incentive regime, insofar as Government used the exportability criterion as its yardstick. Thus, it geared its interventions to international efficiency concerns, and, at least in the 1960s, operated in a relatively sector-neutral fashion. This clearly changed during the HCI episode of the 1970s when industrial intervention was targeted. While the latter approach had dynamic benefits, it left some scars on the Korean industrial scene. Among these were: (i) a troubled capital market, dependent in one way or another on public support despite efforts to set it free of its public sector moorings; (ii) a highly indebted corporate sector, with large credit requirements; and (iii) a highly concentrated industrial structure, dominated by conglomerates, which, were it not for government administrative interventions, would exercise hammerlock-control over credit. Despite the constraints, Government has moved vigorously to redirect its role to functional rather than industry-specific interventions, attempting to signal to industry through its support for the new non-bank financial institutions (NBFIs) as well as through import liberalization that market decision-making will be given preeminence. This emerging approach to the future conduct of industrial policy is seen in recent legislative initiatives as well as in other public sector pronouncements.

5.12 Recent Initiatives. With the increasing sophistication of the economy, the continuing pressure of international competition, and the crises generated by structurally declining industries, Government recognized that a new legal and policy framework was needed to streamline industrial policy. The resulting Industrial Development Law, approved by the National Assembly in December 1985, was conceived in part to replace the seven existing promotional laws governing machinery, electronics, textiles, iron and steel, non-ferrous

---

7/ For example, see Westphal (1982) for a discussion of fostering technological mastery by means of selective infant-industry protection.

Table 5.1: ILLUSTRATIVE INDUSTRIAL POLICY OBJECTIVES AND INSTRUMENTS

Problem area	Market imperfection	Alternative policy remedies
		FUNCTIONAL INTERVENTIONS
Inadequate research spending	Externalities from research	Research and development subsidies* Optimal intellectual property rights* Government research Scientific manpower training Subsidies to "hi tech" sectors
Research duplication; excessive or inadequate R&D spending	Oligopolistic competition for a single discovery	Industry-wide coordination of research* Government research*
Inadequate infrastructure	Infrastructure is a public good	Public provision of infrastructure
Inadequate human capital investment	Moral hazard in long-term lending to individuals	Educational and retraining loans* Subsidies to higher education Government educational institutes Incentives to repatriate citizes residing abroad
Inadequate funding of small and new firms	Immature capital markets	Reform to increase breadth of capital markets* Subsidies to venture capital* Forced allocations of bank funds to small and medium firms Requirements for subcontracting
Inadequate domestic findings for large-scale projects	Small domestic risk poors	Borrowing in international capital markets* Government loans on loan guarantees
High price and low output of a product	Market power in product market	Removal of trade barriers* Laws against anti-competitive behavior*
Differential borrowing rates between small and large firms	Market power in capital market	Reform of interest rate regulations* Removal of international financial barriers* Laws against business concentration Laws restricting concentration of bank ownership

Problem area	Market imperfection	Alternative policy remedies
	SELECTIVE INTERVENTION	
Industry-wide learning-by-doing	Inter-firm technological externalities agglomeration economies	Sector-specific subsidy* Protection of domestic markets Input subsidies (capital, utilities, etc.)
Difficulty in exporting to oligopolized world market	Predatory strategies by foreign oligopolists	Protection of domestic markets* Input or production subsidies
Unemployment in declining industry	Unemployment compensation or high moving costs	Adjustment assistance conditioned on moving* Sector/location-specific subsidy
Excessively slow withdrawal from declining industry	Oligopolistic competition to be sole survivor	Capacity reduction cartel*
Potentially disruptive bankruptcy	Immature capital markets	Government-led restructuring, including mergers and financial aid* In the longer run: financial reform*, improved bankruptcy procedures*, and development of private merger and acquisition markets*

* Asterisk indicates most direct policy instrument; other instruments typically involve additional distortions (negative side-effects).

metals, petrochemicals, and shipbuilding (see Table 5.2). The laws, which played a prominent role over the course of last 15 years, in large measure as signals of Government's industrial objectives, have been abolished. The new framework has adopted a functional rather than industry-specific approach, and is focused primarily on the improvement of industrial technology and productivity.[8/] The framework relies on traditional interministerial policy deliberations for the provision of functional incentives as well as 21 new advisory committees.[9/] A major function of the new law is the restatement of procedures to deal with industrial restructuring. Procedures include the designation [10/] by the Ministry of Trade and Industry of industries to be rationalized and concomitant procedures to coordinate the use of financial and fiscal support. The legislation provides policymakers with a great deal of lattitude with respect to policy implementation, however. Thus, the degree of real intervention will be determined in pragmatic fashion as circumstances warrant.

5.13 It is important, however, that the new industrial policy apparatus lacks a mechanism for explicitly "picking winners," particularly since the industrial policy experience of the OECD countries shows clearly that picking winners is often the first stage of a two-step process to create winners. There are two possible interpretations of the omission of a "MITI-like" mechanism in the blueprint for the future conduct of industrial policy. The first is that the experience of the 1970s with respect to selective intervention, the increasing sophistication of the economy, and the weight of analytical arguments against selective policies have made picking winners unfeasible and/or unattractive. Moreover, industrial targeting of emerging industries now encompasses the additional risk of retaliatory trade actions.[11/] A second interpretation is that while circumstances have changed since the 1970s, the intrinsic nexus between Government and business has not, and key sectoral objectives will continue to be set and supported in a quiet

---

8/ Provisions are made for a Basic Industrial Technology Improvement Plan to be implemented by the Ministry of Trade and Industry and covering aspects of techology and manpower development, research activities, and industrial coordination.

9/ The Councils, based on an extension of the consensus-seeking arrangements pursued by Japan, include 6 functional bodies on manufacturing structure, industrial financing, industrial organization, industrial technology, trade structure, and industrial location as well as 15 bodies on specific industries. The Councils, established under MTI leadership, and including a spectrum of business, academic, and public research participants, will be asked to prepare industrial assessments and outlooks and coordinate flows of information.

10/ This designation is important, for example, in order to qualify for the recently enacted provisions of the Tax Exemption and Reduction Law which allow rationalized industries to sell off holdings exempt from capital gains and transfer taxes.

11/ See, for example, sections of the US bipartisan draft Senate Omnibus Trade Bill (1985).

Table 5.2: MAJOR CONTENTS OF SEVEN MAJOR PROMOTIONAL LAWS

	Industry						
Major contents (year of enactment)	Machinery (1967)	Electronics (1981)	Fabric (1979)	Iron & steel (1970)	Non-ferrous metal (1971)	Petro-chemical (1970)	Shipbuilding (1967)
PROMOTION							
Regulation of Incorporation	X	X	X	X	X	X	X
Regulation of Facilities							
Setting up facility standards	X						X
Approval and coordination of expansion			X	X			X
Encouraging use of domestic facilities	X	X					
Production Regulations							
Regulation of materials imports				X	X		
Production standards and inspection	X						X
Reporting and Inspection		X	X				X
INDUSTRIAL RATIONALIZATION							
Rationalization Program	X	X	X			X	X
Joint R&D Project		X					
Replacement of Old Facilities			X				
TECHNOLOGICAL ASSISTANCE							
Subsidizing R&D Activities				X	X		
CONTENTS OF ASSISTANCE							
Individual Fund	X	X	X		X		
Financial Assistance	X		X		X		X
Subsidy					X		
Direct subsidization	X						
Reducing public utility rate	X						
Tax Preferences							
Special depreciation	X				X		
Reduction and exemption	X						
Special Industrial Complex	X	X	X			X	
Administrative Assistance							
Facilitating overseas activities		X					
Purchase of raw materials				X			
Assisting materials suppliers				X	X		
Business Association	X	X	X				X

Source: Ministry of Trade and Industry.

manner. This view would hold that merely the transparency of policies and the selection of instruments have really changed.[12/]

5.14 The impression gained from the case studies on electronics and textiles (contained in Volume II of this Report) supports the view that the center of gravity of Korean industrial policy is in fact changing but that change will be gradual. The case studies show clear movement towards less selective intervention and more functionally-based industrial incentives, as is appropriate for Korea at this stage of industrialization. This new orientation is bolstered by the absence of a strong central administrative organ of industrial policy.[13/] In practice, it has resulted in a largely hands-off attitude towards both expansion and development of the electronics industry and recapitalization of the textile industry. That is not to say that Government will not be an active participant in industrial discussions, but Government does see its role increasingly focused on coordination and the dissemination of information and on broad functional support for technology.

5.15 These long-term goals must be balanced, however, vis-a-vis short-run interventions necessitated by institutional weaknesses, such as in the financial sector. Thus, the government has intervened, mostly in cases involving "sunset industries," indicating at the same time that it will not act contrary to market signals and will not artificially prop up losing industries. In this regard, Korea has been almost as ruthless in its interventions as themarket could have been.[14/] At the same time, cases of intervention at the firm level seem to be much less commonplace than before; for example, Government disapproved, but did little to prevent, large competitive investments in integrated circuit manufacturing. Government is also somewhat ambivalent with respect to firm size; efforts are being made to foster small- and medium-sized firms in order to offset the power of conglomerates and in order to gain efficiency and spur on new technologies; yet it is also recognized that

---

12/ It is widely reported, for example, that MTI has begun to support medium-sized firms with technical assistance, some special financing, and the latest market information in a move designed to prevent further concentration while still promoting exports in key sectors.

13/ Important industrial decisions will still require decision-making by the Industrial Policy Committee, chaired by the Deputy Prime Minister (also Minister of the Economic Planning Board) and including the Ministry of Finance and Ministry of Trade and Industry.

14/ Government still has strong leverage over firms, primarily because financial liberalization has not yet produced real banking autonomy. Government did exercise its power, for example, in the case of the troubled conglomerate, Kukje, and in the case of over extended overseas construction companies. The visible hand of Government is also seen at times in preventing further industrial diversification; a case in point is automobiles, where it was argued that excessive competition among Korean producers would harm overall export earnings. Samsung Electronics has reportedly tried to enter the automobile business, in partnership with Chrysler, but has failed to gain Government support for this activity.

Korea's future prospects are inextricably intervined with the fortunes of the large firms (see subsequent section on conglomerates).

5.16 Government is having greater difficulty in avoiding industrial entanglements in the case of declining industries than in the case of emerging industries. This is a common dilemma. There are few examples anywhere of industries being allowed to wither without at least some, usually costly and often ineffective, attempt at resuscitation. The experience of European textiles (see case study) as well as the Japanese approach to declining industries (see Appendix 4A) illustrate this point. If one assumes that Government involvement in some aspects of declining industry management is inevitable, the major concern turns to the efficiency of the policy. Continuous subsidization of uncompetitive industries, such as agriculture in many countries, is a costly exercise in "stonewalling" comparative disadvantage. A one-time rescue plan which serves to turn an industry around and save employment at moderate cost may be in the national interest, but for each such successful for intervention there may be several unwise public rescue efforts. Evidence from the OECD countries point to many more failures than successes, a fact of some importance for future policy.[15/] In this area Korea has to date proven itself at least as pragmatic as other industrial societies, with its major handicap being the immaturity of its financial sector.

5.17 Financial Issues. Korea's financial system is still largely based on bank credit rather than diversified capital markets, and the price of credit is set by Government. The rationing principles used to allocate credit can in essence form the blueprint for industrial policy. In the past, Government provided clear signals about its allocational objectives and government regulations took on great importance. This policy is being changed, and government regulations are to be ultimately withdrawn. Since it takes time to develop institutions and financial infrastructure, however, the pace of industrial policy reform will in many respects be determined by the pace of financial liberalization, as discussed in Chapter 4. Although Government may wish to emphasize functional incentives, distortions in the capital market will continue to affect industry-specific outcomes because of the reliance of different industrial sectors on the formal and informal segments of the market. These factors are all the more significant because of the highly leveraged nature of industry. Thus, abrupt Government withdrawal from industrial and financial decisions would not produce neutrality and is probably ill-advised.[16/]

5.18 Constraints on Korean policy as well as reform experiences elsewhere point to the advisability of step-by-step, coordinated policies of liberalization and gradual government disengagement. During the earlier phases of its industrial policy, as reviewed in Chapter 2, Korea managed to integrate the impact of various trade, financial, and industrial policy incentives into a reasonably neutral overall incentive structure. Generally, these policies

---

15/ See Diebold (1980), among others.

16/ Elaborate "second-best" arguments can clearly be made, but on an intuitive level suffice it to say that the rapid decontrol of immature markets is potentially quite dangerous.

have involved compensating or offsetting incentives rather explicit neutrality. The noteworthy trend is that individual elements of policy are now consistently market-enhancing on all fronts. Trade liberalization is moving at an active and consistent pace. Industrial policy is moving toward an indirect, functional emphasis. And financial reform, perhaps the most difficult, has progressed through an initial stage, and, by some accounts, is now simply "stopping to catch its breath" before proceeding in a deliberate manner.

5.19 Risk Management. In Korea, Government has implicitly assumed a substantial share of the risks involved in development. As already discussed, export promotion and later the HCI investment drive relied intensively on preferential access to subsidized (and rationed) credit. In administering this policy, banks became subordinated to government decision-making and the development of financial intermediation lagged. With access to subsidized credit, and under conditions of rapid, large-scale growth, it was not profitable or feasible to attract equity capital. In addition, implicit in gaining access to preferential credit was the notion that Government stood behind the firm as risk partner. Thus, low-cost capital and the implicit socialization of private sector risk led to large increases in indebtedness. In manufacturing, for example, debt-equity ratios stood at about 300% during 1972-74, but rose steady during the decade to reach almost 500% by 1980.[17/] While these levels of indebtedness are not terribly dissimilar from those recorded in Japan in the early 1970s--indeed the current level (1984) of 340% is quite similar to that of Japan--the implications for public policy are far more significant in Korea in light of the differences in underlying economic and financial structure (see Table 5.3).

5.20 High leverage has increased the risk faced by of individual firms as well as the economy's financial structure. Market setbacks of short duration and limited scale can threaten the survival of even large firms--unless credit continues to be provided. Since government policy has sought to minimize the incidence of bankruptcy, partly because of the effects of business failures on the cost of external credit, it has exerted pressures on commercial banks to rescue troubled firms. The risky structure of the economy has, paradoxically, led to an illusion of security because of the perception that the government guaranteed the liabilities of all commercial banks and major non-bank borrowers. Government intervention in financial crises is now expected by borrowers and lenders (at home and abroad) and thus it affects their behavior. Under implicit guarantees, borrowers undertake riskier projects and lenders lend with less careful evaluation of project risk. Large investments in shipping capacity and semiconductor manufacturing are thought by some to be examples of risky borrowing under the implicit government guarantee. In the absence of normal capital market safeguards, for instance, Government has had to use regulatory or persuasive powers to induce firms to strengthen their

---

17/ Financial Statements Analysis, BOK, 1985.

Table 5.3: A COMPARISON OF KOREA AND JAPAN (1984)

	Korea	Japan
Basic Economic Indicators		
GNP ($ billion)	81.1	1,232.7
Population (million)	40.58	120.02
Per capita GNP ($)	1,998	10,271
National saving rate (%)	27.4	30.8/a
Openness Indicators		
Export/GNP (%)	38.5	17.1
Import/GNP (%)	38.1	14.1
External sector/GNP (%)	76.6	31.2
Trade balance ($ billion)	1.4	33.7
Industrial Sector Indicators		
Manufacturing/GDP (%)	28.4	28.6/a
Correlation coefficient between revealed comparative advantage and skill index	-0.51/b	0.18/b
Labor productivity growth in manufacturing (%)	5.7	9.5
Wage growth rate (%)	8.4	4.2
Financial Sector Indicators		
Size of stock market/GDP (%)	7.7	48.0/c
Normal profit ratio (%)	3.41	4.10/c
Average cost of borrowing (%)	14.4	8.0/c
Debt ratio (%)	342.7	324.0/c

/a 1982.
/b 1980.
/c 1983.

financial structures, with only limited success.[18/] Equity markets are still rudimentary and foreign direct investment continues to be limited. Largely absent from Korean industrial finance, with the exception of certain special lending instruments, has been the provision of venture capital.[19/]

5.21 Recently, Government has taken steps to encourage the development of a venture capital market and to broaden industrial risk-sharing. Since existing credit sources are to a large extent tied up in current risk exposures, which cannot be unloaded, the government's strategy is to promote alternative vehicles for risk-sharing in new industrial ventures. New legislation is being considered to offer numerous tax incentives for the establishment of venture capital companies and the mobilization of funds to be invested in them. In the meantime, Government has authorized the creation of a first venture capital trust to channel risk capital to new firms, particularly those in the technology field.[20/]

5.22 There is some parallelism in the approach taken to establish non-bank financial institutions and plans to establish new venture companies, namely, the Korean penchant for creating anew rather than fixing the old. The strategy has attractive features, but some of the problems, for example related to the disproportionate amount of risk being borne by debt in existing companies, are unlikely to be satisfactorily remedied by newly created equity programs for new companies. In particular, neither banks, even if they change borrowers at the upper end of the band of interest rates, nor NBFIs, which implicitly are underpricing capital because they feel protected by the formal banking sector's exposure in firms, are being adequately compensated for risk. Moreover, unless priority attention is given to bolstering the defenses

---

18/ Debt ratios have fallen overall since 1980. In manufacturing, for example, the average debt equity ratio has fallen from 4.9:1 to 3.4:1 over the ensuing five years, but the risks have also become more concentrated in some cases.

19/ In essence, the curb market substituted for a venture capital market; however, with a recent diminution in market segmentation, the size of this market is reportedly shrinking.

20/ Specifically, the approach being considered would offer new sources of finance for the establishment of small- and medium-sized businesses. These incentives would include temporary exemptions from taxes and fees for small, rural or new venture capital companies as well as expedited procedures for the establishment of new firms (the objective being "one-stop" service for the registration of new firms). Government is reportedly willing to commit 50 billion won initially as seed money for certain venture capital parent companies which would then be authorized to create investment companies on a leveraged basis. These new companies could take in investments from the public, with a source of funds disclosure waiver, as well as attract foreign investment. Investment positions could be taken in individual firms or groups of firms, essentially on a mutual investment fund basis. The objective is to attract new equity funds and channel them to promising new industries, especially in the applied technology field.

of firms and industries to withstand export shocks, the government will continuously find itself thrust into market-place to spearhead rescue efforts.

B. Declining-Industry Policy

Introduction

5.23 One of the prices of exceptionally rapid industrialization is the emergence of declining industries, reflective normally of changes in international comparative advantage. These shifts in competitiveness can either be resisted or accommodated. Unfortunately the legacy of declining industry policy in the industrialized countries, which could be expected to provide the template for newly industrializing countries like Korea, is cluttered with examples of resistance to industrial change.[21/] In most cases, the overriding concern which prods public intervention is the maintenance of employment. Fortunately, Korea, as an export-driven economy, has traditionally used international competitiveness as its efficiency yardstick, and its current commitment to both import liberalization and curtailed industrial intervention provide a solid base on which to design an efficient and practical declining industry policy.

5.24 The key to an effective declining industry policy is a market-driven exit policy. There are legitimate economic reasons for public sector interventions in "sunset industries," as more fully described in the chapter Appendix on the theory of industrial policy; these pertain, for example, to the easing of dislocations to labor (i.e., based on distributional or equity considerations) or the reducing of costs imposed on interlinked industries (i.e., involving externalities). However, the basic requirement is that resource allocation decisions be based on market signals and that resources be withdrawn from internationally uncompetitive or domestically non-economic industries. At the firm level, this process begins with the creditor. In the Korean context, the creditors are usually commercial banks, which until recently were government-controlled and which still have not established their technical or managerial autonomy.

5.25 The declining industry arena will prove to be an important testing ground for Government's newly espoused policy of greater industrial neutrality. Procedures will have to be formulated to deal with industrial restructuring in a way which protects the national interest without seriously compromising principles of market efficiency. Moreover, the manner in which these problems are handled will have important ramifications for conglomerate policy and for emerging industry policy.

The Current Situation

5.26 The Facts. Financial difficulties in sizeable industries pose a new challenge to Korean economic policymakers. The problem has taken on major proportions over the last few years as the global economic slowdown has hit particular industries quite ferociously. The slowdown in the shipbuilding industry has been particularly dramatic, for example, as new foreign orders

---

21/ See Adams and Klein (1983) and Diebold (1980) for examples.

plummeted from 3.74 million tons in 1983 to 2.29 million tons in 1984 to 0.64 million tons for the first 10 months of 1985. It is now estimated that the four major shipbuilders will run at 70-85% of capacity in 1986 and in 1987, and avoid a sizeable and prolonged trough in an admittedly cyclical industry, they will have to rely heavily on domestic orders. A related industry, shipping, has suffered from large-scale overcapacity worldwide, leading to record international rates of scrapping. Korean shipping has not been profitable for the past five years and, due to heavy indebtedness, has required a series of government-sponsored financial relief measures. Another industry in considerable difficulty is overseas construction, where new orders have dropped precipitously from over $13 billion in 1981 and 1982 to half that amount in 1984 and less than $5 billion in 1985.[22] Employment in overseas construction has fallen by 70,000 workers between 1982 and 1985, exacerbating an emerging unemployment problem, and the industry cannot expect to recover in light of declining Middle East oil revenues.

5.27 The standard arguments regarding government involvement in troubled industries are complicated by special features of the Korean context. In several troubled industries, particularly shipping and overseas construction, Government intervened aggressively in the upside of the cycle, by providing fiscal incentives and subsidized credit, and by encouraging new entrants into the respective industries. These policies put Government in the position of implicit risk partner and created the expectation that government would provide a "soft-landing" to individual firms should the situation deteriorate. In the case of shipping, for example, the Shipping Promotion Act resulted in a rapid increase in the number of firms (from 19 in 1970 to 64 in 1980) in the liner and bulker trade as shippers enjoyed easy access to bank credit as a favored industry. Similarly, overseas construction was actively promoted by Government during the Middle East's economic boom.

5.28 Although the policy was successful in cushioning Korea's trade balance during the first oil shock and part of the second, the Construction Industry Promotion Law and other incentives led to an explosion of firms authorized to do overseas business. As the Middle Eastern market softened, the industry's highly leveraged firms engaged in sharp price cutting and underbidding to obtain advance payments, and generally lowered the profitability of construction exports. There are striking similarities in the behavior of the shipping and overseas construction industries. Both industries overexpanded, due largely to explicit government promotion polices, and once markets began to weaken, financially strapped firms engaged in destructive competition and widespread distress financing. Ultimately both industries were bailed out by government rationalization programs.

5.29 Rationalization programs in the Korean industrial context typically reduce the number of firms, according to plans carefully orchestrated by Government. Troubled or in some cases bankrupt firm are "placed" with financially healthier ones and, as inducement, Government facilitates the rescheduling of the troubled firm's debts as well as infusions of new credit. In

---

22/ Moreover, a very large proportion of the value of outstanding contracts is concentrated in one or two very large projects, including a $4 billion project in Libya.

heavy electrical machinery, for example, two separate phases of reorganization were needed to arrive at a stable structure. As part of the 1985 Shipping Rationalization Program, the Government reorganized the industry, determining the number of surviving firms, and setting capacity reduction targets.[23/] In both shipping and overseas construction,[24/] the Government strongly encouraged the formation of new "groups" consisting of some strong and several weaker companies, and threatened to withhold valuable financial incentives from firms that refused to cooperate. These restructuring programs mimic, in some respects, the results produced by bankruptcies or takeovers in countries with deeper capital markets. The facilities of a troubled firm are placed under new management, some capacity reductions are ordered, and access to credit is reestablished. The major difference, however, is that Government implicitly accepts some part of the losses of the troubled enterprise, in large measure to prevent undue injury to creditors, i.e., the domestic banking system.[25/]

5.30 Public Policy. The short-term benefits of Korean restructuring operations are clear and substantial. Dislocations are minimized in labor as well as capital markets. Confidence in the Korean industrial and financial environment is maintained. The former is important in the domestic political context and the latter in preserving Korea's access to international financial markets. At the same time, new managerial and financial resources are made available. Yet, Korea's public restructuring programs often fail to impose the discipline that would be brought to bear by market solutions to restructuring. First, lenders are generally excluded from the substantive aspects of the restructuring process: with respect to past loans, their losses are to some extent assumed by the government and with respect to new loans, they are not assigned a central role in either shaping or monitoring the restructuring program. Second, the surviving firms receive mixed signals about the environment in which they must now operate: on one hand, they have an opportunity to

---

23/ The number of firms was reduced from 60 to 15, for example, with two firms opting not to participate in the restructuring. See case study of shipping (Volume II of this report) for details.

24/ A government rationalization program in mid-1984 forced many companies to withdraw from international contract bidding and forced others to cede management, and often ownership, to other stronger firms. The number of firms was reduced from 104 to 49. Thus three of the top 15 overseas construction firms were "placed" with major conglomerates. Further adjustments are likely inasmuch as it is reported that close to W 4 trillion in accounts receivable were due to Korean construction companies in mid-1985 and a very large share of these receivables are in turn due to the domestic banking system.

25/ It is estimated that the banking sector currently holds $4-5 billion in essentially nonperforming loans, most of it in the shipping and overseas construction industries. As a consequence, a program of low-interest loans to commercial banks was announced in July 1985 which would provide 3% BOK loans to commercial banks to enable them to roll over uncollectible assets and preserve their own balance sheet positions. It was widely reported that 1.5 trillion won ($1.9 billion) was to be made available initially.

take necessary measures to reduce capacity [26] and improve long-run profitability, but on the other hand, they have some ill-defined commitment from the government for continuing support.[27]

5.31 In general, the government's willingness to manage restructuring creates significant "moral hazard." Given the prospect of government rescue in the event of adverse business conditions, firms are more willing to undertake risky strategies. Banks may finance such strategies partly because their own exposure is limited by the prospect of government rescue. Once adverse conditions materialize, firms and lenders may postpone adjustment in anticipation of government intervention, engaging in what is sometimes termed "distress finance." Finally, after a restructuring plan is adopted, firms may be reluctant to reduce capacity if they believe that they may outlast other firms, and that in any case they will be rescued again if they fail.

5.32 The feasibility of purely private solutions is to be sure somewhat restricted in the present Korean context. First, the problem of financial distress is not isolated, due to the high leverage of Korean firms and the economy's stake in industries which experienced adverse market changes in the 1980s. Second, the institution of bankruptcy is not well charted; with the exception of certain "demonstration" cases, few large firms have gone bankrupt in recent years.[28] Third, the high leverage of Korean firms makes it difficult for most firms to take over another firm of significant size. Thus, mergers cannot be expected to occur without government financial support. Because of these constraints, market solutions would surely create much greater financial disruption than is the case in economies with more mature financial markets.

5.33 In this tightly-constrained environment, Korean restructuring operations appear to have followed a restrained, middle course. For the most part, restructuring has rearranged management, but left decision-making in the hands of private agents. Second, Korean restructuring plans have typically maintained some competition, to the extent that an industry was large enough to permit the survival of a significant number of firms. Third, direct government intervention has in general been restricted to industry-wide restructuring rather than firm level restructuring, although there are exceptions, such as the restructuring of the Kukje conglomerate and certain large overseas construction companies which were seen as committing the nation to

---

26/ Such severe capacity reductions were effected in the fertilizer and plywood industries, for example, under government directive.

27/ This dilemma is often exacerbated by the fact that stronger firms are saddled with weaker ones, often rendering the new entity financially more vulnerable and more dependent on the financial relief measures being offered.

28/ Even in countries where restructuring is normally handled by market institutions, such as Germany or the United States, financial crises involving entire industries or very large firms (for example, the steel industry in Germany, and Chrysler, Penn Central, and Lockheed in the United States) typically involve the government.

overseas projects. Finally, government has not permitted restructuring to become a routine bureaucratic decision; it has signalled the sensitivity of the problem by requiring that restructuring operations be approved by a small committee at the ministerial level.

Future Policy

5.34 Government policy towards declining industries cannot easily be separated from overall public policy towards the private sector, particularly large firms. Government has begun to issue a revised set of industrial signals. Beginning with the bankruptcy of the sixth largest conglomerate, Kukje, in 1985, public policy has begun to tilt toward reducing the government's role in industrial crises, but it will take more, and clearer examples to alter expectations which have been decades in the making. Public intervention has been necessitated in part by the general weakness of financial institutions and by the absence of strong private mechanisms for industrial restructuring. Thus, clear priorities of Government, if it is to avoid continual industrial involvement, are to: (i) strengthen financial markets; (ii) allow industrial decisions to follow industrial market signals as much as possible, and (iii) limit its own involvement to market failures.

5.35 Financial Issues. A basic shift in financial conditions must take place if the burden of restructuring declining firms is to be shifted in reality from the public to the private sector. In an attempt to promote such a change, Government enacted legislation [29/] which, among other objectives, includes incentives for firms being rationalized or banks involved in the rationalization process to sell fixed assets.[30/] The aim of these incentives is to induce firms to sell peripheral assets, such as real estate, and refinance themselves or help banks to dispose of collateral assets they now hold. Ultimately, these procedures, although somewhat ad hoc, can strengthen the hands of banks and, to some extent, promote greater corporate self-reliance. The procedures involved in this law are still geared to publicly-led restructurings, however, so that government will continue to have to take a central role, at least for the time being, in financial workouts.

5.36 Korea's anticipated industrial transformation will necessitate the withdrawal of resources from certain industries. Between 1983 and the year 2000, for example, Korean planners anticipate a sharp reduction in the output shares in petroleum, textiles, and food and beverages, among manufactures, as reported in Chapter 1, Volume II. Thus exit policy procedures which work more smoothly than at present will be needed, and the role of the banking sector, as the prime creditor, will need to be strengthened. Inasmuch as other lenders, such as the NBFIs, base their lending decisions, especially vis-a-vis large corporate borrowers, in part on the relationship between the borrower

---

29/ The Tax Reduction and Exemption Law passed on December 18, 1985.

30/ Firms may either exempt 6% of the value of the transferred asset or the depreciation allowance on the transferred asset may be taken in subsequent years as a special expense and loss. Banks will be exempt from the special value-added tax applied to marginal profits (viz. capital gains) or properties acquired via rationalization programs.

and its prime bank, commercial banks will eventually have to take responsibility for financial workouts. This argues for greater real bank autonomy, which can only be established on the basis of stronger bank finances, managerial autonomy from both the government and its major borrowers, and a revised public yardstick for judging bank performance--i.e., profitability.

5.37 Industrial decision-making in Korea should in any case be refocussed to stress the primacy of profits. Clearly this is the ultimate market test of efficiency and entrepreneurship, yet it has often been subordinated in Korea in the name of industrial growth. Hence the existence of very large, negligibly profitable firms, hand-pressed to raise equity capital. Much of the Korean corporate penchant for "growth at all costs" is interwoven with historical public policies, such as the allocation of credit to the largest or fastest-growing firms. But, in order for market agents to make rational decisions concerning the viability or future prospects of declining industries, for example, a clearer pursuit and reporting of profits is important. This is a complicated issue in the setting of Korean conglomerates, insofar as subsidiary and parent profit maximization do not necessarily yield identical industrial strategies. The banking sector, for its part, should be clear whether it is lending to the firm or the conglomerate, and on the basis of whose financial position declining industry decisions will be made.

5.38 Industrial Signals. While Government has begun the process of altering industrial expectations, it must continue to: (i) press ahead with trade liberalization to provide correct efficiency prices by which domestic producers can judge their competitiveness; (ii) let market prices determine the ultimate viability of industries and refrain as much as possible from intervention; and (iii) indicate both the strictly limited circumstances under which it would intervene (see Box 5.1 on decision rules for intervention) and the cost to the nation of that intervention. In the last instance, the intervention with implicit protection as practiced by some industrialized countries can be shown to be costly for the Korean economy in terms of both efficiency and trade diplomacy (see Appendix 4A of Volume II on Japanese Restructuring Policy and its limited application for Korean policy). The longer that market signals can be ignored--for example by using implicit government support for the commercial banking creditors as a crutch--the costlier the ultimate adjustments will turn out to be. In this instance, creditors need to form earlier judgments on the potential for reversibility of decline, an issue clearly related to the risk-return calculus of banks and bank profitability.

5.39 The government has by virtue of its direct involvement in industry level restructuring exercised a strong influence over the resulting structure of the industry. In some cases, like shipping where national security concerns abound, Government has dictated the ultimate number of surviving firms and has designated merger partners. In overseas construction, the government has also materially affected the industrial outcome by seeking to force mergers, essentially in the form of a strong firm absorbing a weak partner, and ensuring that overseas contracts are completed and that Korea's reputation abroad is not tarnished. Much of Government's role as industrial organizer stems from its de facto role in industrial finance and the weaknesses of the financial sector. In the future, as financial markets mature, Government should limit the use of its "good offices" to industry-wide restructuring efforts (rather than firm level restructuring) in exceptional cases. Only in the unique case where competitors cannot agree on capacity reductions in industries with large capital investments and scale economies or in industries which are competing in oligopolistic international markets should it actively engage in industry production decisions. Essentially, the public sector has no advantage over markets in determining industrial structure, and the prevalence of market failures is not pervasive enough to warrant intervention unless anti-trust considerations are overriding.

5.40 The issue of trade liberalization merits re-examination in the context of declining industry policy because international prices provide the best signals of both report competitivenes and domestic viability of industry. Import prices are thus valuable industrial guideposts. It can be expected, however, that a troubled industry selling in its domestic market will request protection from foreign producers. This story has been repeated in virtually all the industrialized countries in textiles (see case study) and has resulted in considerable trade barriers. In some cases, the industry losing its external comparative advantage has opted to protect its domestic rents by jointly reducing capacity with other producers (and receiving anti-trust waivers from government). This was the case in Japan as described in Appendix 4A of Volume II. This approach can only succeed behind some form of protection, however, and generalized protection to help maintain a declining industry's domestic market share can be counterprotective if it damages the country's export prospects. Thus, Korea should continue to heed international price signals of competitiveness for its import-competing industries and continue to use safeguard mechanisms sparingly.

5.41 Other Government Roles. Dislocations in labor markets as a result of declining industry developments are often the most troubling aspect for governments. Retraining programs may have to be devised to ease labor adjustments and deal with humanitarian aspects of redundancy. Still, since the Korean employment situation is relatively free of many of the market distortions extant in the industrialized countries, its labor market adjustments as a result of industrial change may be more manageable than elsewhere, and thus, the efficiency of industrial restructuring (at least on the labor market side) may preclude the need for major interventions.

5.42 Government considers itself, and rightly so, the guardian of the nation's image, a public good of some consequence. In an effort to protect that image, the Korean government is willing to go to extraordinary lengths. Clearly this voluntary extension of the concept of sovereign risk to private transactions has enabled Korea firms to borrow more cheaply on international capital markets and may have helped them to win international contracts. As industrial transformation continues, however, commercial losses will occur. And it would be ill-advised for Government to shoulder these losses and repay private sector debts or fulfill private sector contracts. Korea is at a stage of development where its private sector should be distanced from its public sector in terms of risk-bearing. This is a necessary and healthy development insofar as it conserves public resources, but more importantly because it forces foreign firms and banks to begin to assess Korean corporate risk more selectively, and, insodoing, it may hasten the process of risk differentiation in the Korean financial sector.

Box 5.1: Possible Decision Rules for Intervention

Policymakers faced with requests for assistance from troubled industries need to follow predetermined steps with respect to public sector involvement. While there will always be room for "enlightened pragmatism" in the end, the initial public policy response should be structured and predictable. Insofar as Korea's industrial structure is fast-moving and the markets in which it competes somewhat unpredictable, it is reasonable to assume that declines will continue to occur. Private sector decision-making and planning would be improved if Government dealt with these industrial crises on a systematic rather than ad hoc basis. The decision process might involve the following series of steps: (a) an assessment of the direct and indirect costs to society of the distressed industry's performance, that is, a judgment of the extent of externalities (the "intervention test");/a (b) an analysis of the source of decline and a judgment on its reversibility (the "reversibility test"); and (c) if a public solution is needed, realistic near-term and medium-term objectives should be established, based on the source of the decline, appropriate policy tools should be selected to bring about public objectives, and limits should be placed on both the duration and cost of intervention ("the efficiency test"). The tools selected should provide incentives to private agents to pursue efficient solutions, i.e., to merge or shut down capacity if circumstances warrant; public restructuring should neither reward past or current uneconomic behavior. The intervention should be monitored vis-a-vis its public objectives ("the monitorability test") and potentially reversed or abandoned if it fails to measure up.

The first step is important because the source of decline will reveal whether internal management or external forces are the major determinant of difficulty, and, therefore, whether changes in the management/ownership of resources are needed or whether the allocation of resources to the industry is to be questioned. In judging the adequacy of management, lenders should take on an increasing role, although this may be complicated by interlocking directorate problems. In cases where genuine externalities seem to exist, i.e., where public policies may need to be amended, a more direct public sector plan may be necessary and this may require a public sector view on the reversibility of the decline. Such judgments should be based as much as possible on narrowly defined economic criteria. Finally, if direct public intervention is warranted, it is important that the "dosage be consistent with the disease" and policy interventions be time-limited and dependent on certain corporate actions, i.e., conditional. The test is that every marginal expenditure of public resources committed to a declining industry must pay for itself in social benefits.

/a A presumption would exist for a private solution, either restructuring, merger or outright dissolution unless the dislocations and externalities were overriding or there were no viable "workout agents." The objective should clearly be to withdraw resources either from inefficient entrepreneurs or from inefficient activities entirely. If a market-based solution is feasible, and in many cases as noted earlier the absence of functioning markets (such as for capital) may make this difficult, Government will need to ensure that supporting policies are in place to underpin the workout and provide assistance (vis-a-vis employment, for example).

5.43 In general, policy procedures toward restructuring are still embryonic, and the degree of future intervention is not yet known. A case-by-case approach to restructuring can maximize the flexibility of policymakers but provides murky signals to the private sector. Insofar as Korea's industrial structure will be subject to many more jolts in the future, both as a consequence of rapid changes in comparative advantage and as a consequence of corporate miscalculations, there is merit in establishing more explicit intervention criteria and avoiding ad hoc policy measures

C. Conglomerate Policy

The Current Situation

5.44 Korea's industrial development is intimately intertwined with the emerging economic power of Korean conglomerates, the jaebol.[31/] Recent data on Korean manufacturing, as reported in Chapter 1, Volume II, reveal that the top 50 firms account for more than one third of output, a concentration considerably higher than that of either Japan or Korea's major competitor. The structure of industry is very much a product of government policy to initially encourage industry entrants with a minimum equity stake, encourage competition, and then select future industry leaders, essentially following a "survival of the fastest growing" criterion, and to provide those firms with preferential credit. The resulting concentration of industrial production is evident even below the largest 50 or 100 firms; as of 1983, for example, the largest 1,000 firms (on the basis of employment) accounted for 63% of national value-added and 45% of employment. On the trade side, concentration is even more pronounced as the nine officially designated General Trading Companies account for almost 50% of Korean exports.

5.45 There have been clear benefits for the Korean economy of industrial concentration, among them the gains associated with scale economies, name recognition in foreign capital markets, and product diversification. Moreover, concentration has made coordination between the public and private sectors more manageable. Economic agents have been easier to control, their efforts more quickly marshalled, and their positions strengthened in international competition by virtue of size. The HCI episode of the 1970s served to reinforce the bias toward "bigness" by fostering the development of several strategic industries in which scale and therefore limited domestic competition were essential, While the end result of the incentive structure which fostered expansion, diversification, and mergers was close to a dozen highly competitive firms by world-class standards (viz., 10 of the 27 developing country firms in the Fortune 500 category are Korean), it also produced cracks in the Korean industrial system.

5.46 There appear to be three major dangers with respect to heavy economic concentration. First, there may well be efficiency losses associated with very large firms. Second, the economy may be more vulnerable to financial shocks if economic activity is highly concentrated, although it must be

---

31/ See Jones and Sakong (1980) for details on historical development.

acknowledged that conglomerates are by their very nature more diversified. Third, economic management may prove more difficult as the relationship between Government and the key private sector entities becomes more complicated. Realizing the potential difficulties of economic concentration, particularly at the very top, the government has attempted to promote small and medium industries (SMIs). Policies have included reservation of a predetermined share of credit for SMIs, and recently, new tax and other incentives to promote the establishment of new firms. At the same time, there have some efforts to contain the spread of conglomerates by trying to limit their credit access and force divestiture of peripheral holdings to shore up the finances of subsidiaries operating in the "main line of business."

5.47 Financing Issues. Underlying some of the concern about conglomerates is their thin capitalization and extremely diversified nature. While diversification is normally associated with reduced risk, Korean conglomerates are highly leveraged and operate in product markets characterized by highly cyclical demand (shipbuilding), protectionism (textiles, footwear, possibly autos), and cut-throat competition (shipping, semiconductors). Vulnerability is therefore high and Government becomes an automatic risk partner, in light of either past involvement in industrial decision-making, perceived public diseconomies of failure (related to either domestic finances or foreign perceptions of Korea's creditworthiness), or actual public externalities (such as employment).

5.48 There are risks associated with the pervasive access to credit on the part of conglomerates. Conglomerates can borrow scarce (and implicitly rationed) credit on the basis of their masthead rather than on the basis of the end use of that capital. While an argument can be made that in a distorted credit system, the efficient allocation of capital inside a conglomerate has certain second-best benefits, this is a tricky argument. One needs to consider that Korean jaebol are driven first by size and therefore growth, second by wingspan (namely the number of critical areas in which they hold dominant positions), and perhaps only third by profits. Therefore even internal jaebol capital allocations may be biased. This situation is exacerbated by the highly leveraged feature of Korean industry which produces many thinly capitalized subsidiaries, none of which can be allowed to fall for fear of damaging the overall conglomerate name and ultimately access to capital.

5.49 More broadly, the government's view has been somewhat ambivalent vis-a-vis conglomerates, since they pose perhaps the best chance for the desired industrial leap to mature country status but also embody large risks, and, as has been argued, these are likely to become sovereign risk in financial crises.[32/] While "camping with one's back to the river,"[33/] is

---

32/ It should be added that even international lenders have been taken in by moral hazard behavior insofar as lending continues to known high corporate risk firms because lenders firmly believe that any debt will be treated as sovereign debt in a crisis.

33/ Korean proverb indicating that cutting off retreat, while riskier, forces a harder fight.

nothing new for Koreans, the economic grip of the conglomerates and their high stakes strategy is making policymakers increasingly uncomfortable.

5.50 Policy Choices. Since conglomerates are in a position of influencing outcomes in both factor and product markets, and due to the public externalities of their actions in terms of Korea's international borrowing and international marketing, Government has an obligation to examine their operations and act in the public interest. Issues of course exist with respect to when and how public actions should take place. Korean policymakers face the difficult choice, for example, of either: (i) allowing highly leveraged and highly diversified conglomerates to provide the locomotive for the economy's push to mature country status, and essentially playing a largely accommodating role, which involves accepting the attendant risk of a derailing due to either financial overreaching or business miscalculation or (ii) reining in the jaebol, forcing them to raise more equity to satisfy creditors and implicitly be more concerned with profitability, but by doing so perhaps lose critical time in Korea's development progress. Ironically, policies which curtail jaebol expansionism in hopes of reducing the extent of nationally-borne risk may serve concomitantly to raise the risk of derailment. It is fairly clear in light of Korea's history and national economic temperament that it will choose to follow the highest return strategy, accepting the attendant risks.

5.51 Faced with these realities, the Government has indicated two important objectives of policy. The first relates to the need to maintain a balance of power necessary for a concensus-style industrial state, a continuing objective of policy. Hence, the active promotion of small and medium industries. The second relates to the need to institute economic safeguards to reduce societal risks associated with conglomerate actions. Put differently, Government, and implicitly the public, could afford to take a risk partnership position with firms in the 1960s but the HCI episode showed very clearly the costliness of that approach. While Government removed itself from direct risk partnership (via industrial diverstiture, for example), it is still implicitly a risk partner to some extent. Recent financial difficulties and ensuing bank vulnerability demonstrate that the level of public risk-sharing needs to be reduced. Hence the nexus between conglomerate policy and domestic financial reform. Issues related to SMI and conglomerate policy are identified in the following section.

Future Issues

5.52 SMI Policy. It has been a clear priority of Government to provide credit preferences for small and medium industries (SMIs) in recent years. This policy has been based on a number of objectives, including an "affirmative action" program to offset past discriminatory access to credit and prevent any abuses of credit markets by monopsoistic borrowers, i.e., the large firms. These measures to overcome distortions in the credit market are in the nature of "second-best inverventions" and should be considered prudent and useful interim measures until capital market biases can be wrung out of the system. In the absence of these credit market distortions, however, Government should be expected to be ultimately more neutral in its stance on the size of production units since there is no conclusive evidence that smaller firms are inherently more efficient.[34/] Moreover, even if smaller production units have some technological or cost advantages, it does not imply that the ultimate ownership structure should favor SMIs unless Government has additional public policy objectives in mind.

5.53 Other objectives of an active SMI policy can include equity considerations, related to income distribution, for example, or general anti-trust goals. It is often argued that to limit the power of the conglomerates, the seeds of new conglomerates need to be spread. Thus, SMIs are seen as a political as well as economic counterweight to the jaebol.[35/] A second rationale for explicit SMI promotion revolves around employment generation inasmuch as smaller firms are more labor-intensive. Indeed, an argument can also be made that smaller firms are better able to withstand demand shocks, since they have greater flexibility to adapt their production techniques and products. This flexibility, is, of course, dependent on access to capital. Finally, in addition to employment externalities, there may be technological advantages to the economy of SMIs in terms of innovation; however, a distinction should be made between small industries and medium industries because some minimum production levels are normally needed to sustain certain modern technological innovation.[36/]

5.54 Bank Ownership. One issue already identified concerns corporate ownership of banks (see Table 4.10). The arguments against major equity holdings by conglomerates which also borrow extensively from those institutions is straight-forward, and based on potential conflicts of interest. Moreover, even though SMIs are allocated a fixed share of bank credit, the firm level decisions can be biased in favor of those SMIs with links to large firms. More broadly, every effort should be made to strengthen banking autonomy, and parallel efforts should be made to bolster the financial structure of conglomerates, i.e., lower their indebtedness. As a general policy, special tax breaks for conglomerates are not preferred public policy, even if their ultimate objective is to shore up bank finaces. A preferred approach would be to alter the ways banks judge their risk exposure and support bank efforts to reduce the riskiness of their portfolios by imposing normal credit conditions on all borrowers, regardless of size.

5.55 If banking reform is successful, second best kinds of policies, such best kinds of policies, such as restricting the spread of conglomerates into new product lines, would be unnecessary. Normal financial tests would be

---

34/ See Chapter 1 of Volume II for productivity comparisons between SMIs and large firms.

35/ This may well account for the aggressive pursuit of vertical integration on part of the conglomerates, wanting to maintain their economic position.

36/ See Westphal, Kim, and Dahlman (1984).

imposed on such activities.[37/] Hence the importance of continuing financial reform to effectuate Government's desired withdrawal from industrial interventions. In the context of domestic markets, antitrust policies may at times be necessary, but their incidence can be minimized by a continued emphasis on import liberalization.

D. Emerging-Industry Policy

Current Setting

5.56 Korea is embarking on a major industrial transformation. It has entered the automobile export market for the first time. It has adopted a high-risk, leapfrog strategy with respect to the technological requirements of the semiconductor industry, as described in detail in the case study on electronics (see Volume II of this report). More broadly, the Korean strategy to develop and utilize technology to the fullest is a departure from past policies which were based on superior organizational ability and the dedicated productivity of labor. Two aspects of this technology-intensive strategy are noteworthy and are borne out by the analysis of the electronics industry: first, the strategy is high risk not only because of the technological sophistication involved, but also because it puts Korea in a head-to-head confrontation with the two industry leaders, Japan and the United States; and second, the strategy is costly in terms of capital requirements and this has implications for Korea's capital markets, especially for risk capital, which are still developing.

5.57 These dual constraints are bound to put policymakers' resolve to the test with respect to the articulated desire of Government to withdraw from explicit involvement in industrial decisions. Certainly, both trade diplomacy and credit access are areas where public intervention has come to be expected in Korea. One of the important future issues, therefore, is the extent to which Government may be drawn into an interventionist posture. This issue is interwoven with the issue of conglomerate policy, discussed previously, inasmuch as it is the jaebol which are at the vanguard of the push into new industries.

Public Policy

5.58 The era of explicit import substitution or directly promoted infant industries appears by most accounts to be generally passe. In the past when market incentives failed to produce new industries, Government either created public enterprises or promoted their private creation. The increasing sophistication of the economy, along with a diminuation of the extent of market failures, has led policymakers to be less inclined towards interventions. Government policy now explicitly favors functional incentives, and policymakers have gone to considerable lengths to stress the reorientation of public policy. This does not mean, however, that Government lacks control

---

37/ These tests would also prevent uneconomic vertical integration attempts by conglomerates.

over the emergence of new industries or that it lacks industrial aims. It does connote that Government is confident enough of the effectiveness of functional incentives and of the announcement effect of its future industrial goals that it no longer feels the overwhelming need for, or is that comfortable with, heavy-handed intervention on behalf of emerging industries.[38/]

5.59 As has been noted in earlier sections, there are implicit linkages between the provision of credit in the formal financial market and industrial developments as well as between conglomerate policy and industrial outcomes. Government thus can exert its influence in a variety of ways. Moreover, Korea has throughout its modern history been quite adept at recognizing, and balancing, the net effects of its policies on industrial development. Thus, public sector oversight will continue, and to the extent that markets are still imperfect, or firm concentration disturbingly high, this is clearly warranted. Explicitly, Government has grounded part of its future industrial role on the basis of information exchanges, although past performance should indicate to all industrial participants that all of the public sector's powers can be brought to bear if needed.

5.60 The important issue with respect to emerging industry policy is how and to what end these explicit and implicit power will be used, which in turn revolves around the concept of industrial "vision." Without delving into the question of whether the public sector's vision is superior to that of the private sector, one can point out the potential pitfalls of visionary industrial policy. The experiences of most industrializing countries, including Japan, reveal that there is no perfect vision. The record on Japan is clear insofar as its success is concerned, but less lucid with respect to the degree of intervention.[39/] Korea has learned from the HCI experience--which was costly in many senses, effective in some, and unsuccessful in others--that massive capital investments guided by Government are not advisable. As it has now turned to technology as the major vehicle of industrial transformation, Korea must now develop a future industries-cum-technology policy.

Future Issues

5.61 Technology Policy. There are strong conceptual reasons for an activist public policy towards certain technological investments. Those arguments favor functional incentives and can be extended to include selective interventions with respect to acquiring information and coordinating

---

38/ There will clearly continue to be exceptions, however. The auto parts industry, for example, has been "targeted" as an important industry with a bright future. Most producers are small- and medium-sized firms which, if properly financed, can support the emerging auto export industry. Government has been inclined to provide industry-specific support, in the form of tax and other incentives.

39/ See C. Johnson (1982) *inter alia*.

investment in the pursuit of dynamic efficiency.[40/] Many of these arguments, to be sure, rely on inter-firm externalities and are likely to have limited applicability in Korea where industry is highly concentrated. To a very significant extent, in fact, large firms finance their own large-scale investments, including technology related and R&D expenditures, without public guidance or explicit support. It is reported, for example, that the four most prominent conglomerates will spend a total of almost $3.5 billion in 1986 on plant and equipment and R&D related investments, approximately 20% of total national private investment. It is noteworthy that the bulk of the large investments in the high-technology field have been privately funded (see case study on electronics).

5.62 An area of technology development in which Government has felt less confident of private initiative--perhaps because of the dearth of risk capital and the greater incidence of interfirm externalities--on the part of small and medium firms. For this reason, new venture capital programs have been authorized, although it is unclear how large an impact these special measures will have in light of the economic size of the conglomerates. Korea is pursuing technology investments with unbridled vigor. No potentially profitible field is being ignored, including biotechnology and genetic engineering. Insofar as technology itself is a production input, much of the potential rewards to be derived from technology will depend on the flexibility of the productions process and of the economy as a whole.

5.63 Industrial Flexibility. It can be argued that the major factor which will influence industry's receptivity to Government's policy initiatives with respect to technology and information is its flexibility. Flexibility is most important on the side of emerging industries, as the economy is expected to shift its structure in response to newly developing comparative advantage. Response speed is imperative in some markets because of the limited window of opportunity for some products and the strong competition from other would-be producers. Furthermore, since the pursuit of dynamic comparative advantage entails risks, it is to be expected that some "advantages" will not materialize in actuality, and in those cases, industries must cut their losses and reallocate resources. For these adjustments to take place, fexibility in industrial structure is equally vital. As a policy matter, therefore, Government may wish to take steps to improve the flexibility of the economy and ensure that new rigidities do not emerge in either factor or product markets. One of Korea's great assets is its flexibile and vibrant brand of entrepreneurship. Firms themselves seek out opportunities and respond vigorously to financial incentives. Government policy can assist this flexible entrepreneurship with appropriate incentives to technical innovation, reduced financial distortions, and a continued emphasis on allocative efficiency.

40/ See Pack and Westphal (1985) for a discussion of the merits of selective intervention in support of an industrial strategy based on managing technological change. See also Westphal, Kim, and Dahlman (1985), on the sequential nature of Korea's acquisition of technological capacity.

5.64 The economy's flexibility is interwoven with its resiliency. The clearest example of this nexus is seen with respect to capital. As new market opportunities arise in which investments, say in technology are needed, capital, i.e., essentially risk capital, needs to be generated, and if that capital is withdrawn from other highly leveraged ventures, it subjects all the enterprises to potentially higher risk. Therefore, especially in the Korean case, the flexibility of the economy can be enhanced by strengthening firm finances and by increasing the mobility of capital. Capital should flow to the highest yielding activities, on a risk-adjusted basis. This process can be strengthened by general financial reform as discussed in Chapter 2 and by a clearer distancing of Government from the risk-taking activities of large firms, particularly the jaebol. While Government has clearly recognized this, the nature of proposed industrial consensus-building which appears to underline recent legislation, may reinforce the perception that the public sector will continue to be the lender of last resort for large corporate entities. This perception may serve to encourage corporate financial behavior which is at odds with the goal of increasing the flexibility of the economy.

5.65 Government may wish to explicitly reduce distortions and rigidities and seek to assist the ability of the industrial sector to finance, develop, and absorb technical innovation. In this respect, public policy which attempts to promote the SMI sector is probably helpful, especially in view of the inherent biases in favor of large firms and their often inefficient policy of vertically integrating at the expense of the SMIs. Technical innovation by efficient production units is important to the future of the economy because it fosters increased returns to factors of production. As noted in Chapter 4, total factor productivity has fallen in Korea in recent years, a trend Korea has in common with advanced countries, but one which is premature in a rapidly industrializing country with high income aspirations.

5.66 Korea has ambitious plans with respect to R&D expenditure and human capital investments, substantially aimed at new industries. For example, ambitious quantitative targets for the proportion of GNP allocated to R&D expenditures have been set, with the ratio rising from 1.7% in 1985 to 2.0% in 1986 and 3.0% in the year 2001. Much of this investment is in basic science and the goals are state-of-the-art technology developments.[41/] Technologies can still be purchased, although there are clear advantages even in securing foreign technology to having some credible indigenous capacity. Nevertheless, unlike Japan and the US, Korea has a limited domestic market for some of the higher technology products and care should be taken not to "put too many eggs in the higher-tech basket." In the area of human capital development, there are projected shortages in higher skill categories, some of which are quite binding in relation to the ambitious R&D targets. Public sector interventions

---

41/ This focus should not blind policymakers to the potentially high returns to innovation in some more traditional export industries, however, where technical advances may be necessary to either preserve markets or prevent their unneccesarily rapid decline. Risky technological investments need to be balanced with more traditional and inherently less risky productivity-enhancing investments.

are clearly warranted and necessary in this area; however, care should be taken not to overemphasize quantitative targets with respect to skill categories.

5.67 Financing. Financial retrenchment is important for the future strength and responsiveness of the economy because it can reduce the vulnerability of industries to external shocks and free up necessary capital, either via profits or borrowing capacity, for new industries. Emerging industries will require substantial capitalization. In allocating that capital among competing users, it will be important for the price of capital to be eventually decontrolled and for capital to be absorbed by the highest yield activities. This means that finacial reform is a priority. Moreover, since emerging industry investments will be inherently risky, it is important that equity holders bear the primary risk, debt-holders the secondary risk, and government bear as little as possible. To attract equity capital, firms engaged in new industrial activities will have to generate profits, an objective perhaps not easily achieveable in emerging industries like autos and semiconductors in which Korea is attempting to establish market presence. In this case, care must be exercised to see that government does not end up underwriting these high-risk forays into new products and markets and into advanced technology.

5.68 One dilemma facing policymakers in this connection is the extent to which Government should erect entry barriers to conglomerates interested in a diversified pursuit of newly emerging industry market shares. One economic reason to limit entry is to maximize international rents captured by the nation in its exporting activities. At the same time, long-term efficiency requires that large firms not be granted market-controlling power in new products or in arrears. Thus, in highly selected cases, a governmental gatekeeping role may be warranted, but in general, it should be lenders who establish entry barriers to industrial diversification based on financial criteria. Therefore, public sector constraints on conglomerates should be a temporary fallback measure activated in light of capital market imperfections.

5.69 Allocative efficiency is an important prerequisite for judging the advisability of new ventures and prospects for developing dynamic comparative advantage. Factor and product market imperfections will increasingly be reduced as fianncial and trade liberalization proceeds, so that price signals should have greater relevance to decision-making. Public policy should aim at reinforcing those market signals rather than managing them. In the final analysis, the effectiveness of public policy depends not only on its aims and its monitorability, but also on the impact it has on the behavior of market agents. Korea has show itself to be pragmatic and flexibile in managing industrial policy in the past. Now, as it begins to withdraw from direct economic management, it will increasingly have to make its policies transparent and predictable. In sum, Government can neither pick, create nor mandate "winners," at least not without significant costs. It can, however, provide an environment in which winners emerge and thrive. This is the challenge of Korea's future industrial policy and the key to determining the speed at which it gains mature country status.

KOREA

STATISTICAL APPENDIX

Page No.

A NOTE ON NATIONAL ACCOUNTS METHODOLOGY

1. The Bank of Korea recently adopted a New System of National Accounts (New SNA) and has revised its national accounts accordingly for 1980-85. As a result, an incongruity exists between national account figures prior to 1980 (based on old SNA) and those for 1980-85. National Accounts data for the post 1980 period will represent new SNA figures unless otherwise noted.

2. Major changes in the New SNA are as follows:

(a) Transactions of goods and services are classified by activity whereas those of income and expenditure are classified by institutional sector;

(b) Government and private nonprofit institutions are classified as independent producers;

(c) Gross intermediate output in addition to final products are estimated;

(d) Imputed financial services and import duties are introduced;

(e) Current value-added are estimated by subtracting intermediate inputs from gross output; and

(f) Double deflation method is used extensively for estimation of current value-added.

3. A comparison of the old and new SNA methodologies as they affect GNP accounting is provided below:

GNP (billion won)	1980	1981	1982	1983	1984
GNP					
New SNA (A)	36,672.3	45,126.2	50,724.6	58,985.8	66,408.2
Old SNA (B)	37,205.0	45,775.1	51,786.6	58,428.4	65,379.7
(A)-(B)	-532.7	-648.9	1,062.0	557.4	1,028.5

GNP and GDP Growth (%)	1981	1982	1983	1984	Average of 1980-84
GNP					
New SNA	6.6	5.4	11.9	8.4	8.1
Old SNA	6.2	5.6	9.5	7.5	7.2
GDP					
New SNA	7.4	5.7	10.9	8.6	8.2
Old SNA	6.9	5.5	9.5	7.9	7.5

Table A1.1: NATIONAL ACCOUNTS SUMMARY /a
(billion won at current prices)

	1980	1981	1982	1983	1984	1985P
Gross Domestic Product	37,914.9	47,023.7	52,912.7	61,002.9	68,866.7	74,977.9
Resource Gap (M-X)	2,963.9	2,520.8	1,351.3	782.2	207.6	-424.4
Imports (G & S)	15,729.3	19,712.3	20,153.6	23,027.9	26,037.2	26,904.1
Exports (G & S)	12,765.4	17,191.5	18,802.3	22,245.7	25,829.6	27,328.5
Total Expenditures	40,843.0	49,560.0	54,621.2	61,656.1	69,064.7	74,327.4
Consumption	29,054.1	35,881.0	40,111.6	44,035.3	47,857.4	51,787.4
General government	4,268.0	5,383.4	6,110.3	6,753.4	7,079.1	7,782.2
Private	24,786.1	30,497.6	34,001.3	37,281.9	40,778.3	44,005.2
Investment	11,788.9	13,679.0	14,509.6	17,620.8	21,207.3	22,540.0
Fixed investment	11,835.7	12,931.0	15,486.5	18,479.6	20,795.0	22,091.0
Change in stocks	-46.8	748.0	-976.9	-858.8	412.3	449.0
Domestic Savings	7,618.2	9,245.2	10,613.0	14,950.5	18,550.8	20,529.6
Net factor income	-1,242.6	-1,897.5	-2,188.1	-2,017.1	-2,458.5	-2,660.9
Current transfers	996.4	1,356.2	1,584.9	1,535.6	1,569.8	1,480.5
National savings	7,372.0	8,703.9	10,009.8	14,469.0	17,662.1	19,349.2
Average Exchange Rates						
Won/$	607.4	681.0	731.1	775.8	806.0	870.0
Won/SDR	790.6	803.0	807.2	829.3	826.2	883.3

P = Preliminary.

/a The following four tables have been prepared according to standardized World Bank concepts and definitions to facilitate cross-country comparisons and aggregations. These data may not always agree with similar data in the main text and this statistical appendix.

Source: Bank of Korea, New System of National Accounts, 1986.
Bank of Korea, Preliminary Estimates of 1985 National Accounts, 1986.

Table A1.2: NATIONAL ACCOUNTS SUMMARY
(US$ million at 1980 prices)

	1980	1981	1982	1983	1984	1985P
Gross Domestic Product	62,418.6	67,042.0	70,848.8	78,559.5	85,396.7	89,796.0
Terms of trade effect	-	97.7	312.9	655.6	1,262.2	1,186.2
Gross domestic income	62,418.6	67,139.7	71,161.7	79,255.1	86,658.9	90,982.2
Resource Gap (5-6)	4,879.4	2,779.3	1,539.8	215.9	-258.0	-1,450.0
Imports (G & S)	25,894.8	27,051.8	27,597.1	30,593.4	33,690.6	33,182.1
Capacity to import	21,015.4	24,272.5	26,057.3	30,377.5	33,948.6	34,632.1
Exports (G & S)	21,015.4	24,174.8	25,744.4	29,721.9	32,686.4	33,446.0
Total Expenditures	67,239.0	70,207.4	72,305.1	79,633.4	87,037.8	90,306.7
Consumption	47,831.2	49,569.8	51,639.0	55,359.0	58,240.9	61,054.1
Government	7,026.3	7,460.8	7,522.0	7,920.1	7,959.8	8,444.6
Private	40,804.9	42,109.0	44,117.0	47,438.9	50,281.1	52,609.5
Investment	19,407.8	20,637.6	20,666.1	24,274.4	28,796.9	29,252.6
Fixed investment	19,484.9	18,700.9	21,105.5	24,722.0	27,357.1	28,126.7
Change in stocks	-77.1	1,936.7	-439.4	-447.6	1,439.8	1,125.9
Domestic Savings	14,528.4	17,858.3	19,126.3	24,058.5	29,054.9	30,702.6
Net factor income	-2,045.7	-2,691.0	-3,003.0	-2,691.0	-3,077.7	-3,246.0
Current transfers	1,640.4	1,855.2	2,164.4	2,039.1	2,039.4	1,831.2
National Savings	14,123.1	17,022.5	18,287.7	23,406.6	28,016.6	29,287.8
Won Deflators						
Gross domestic product	1.00	1.15	1.23	1.28	1.33	1.37
Imports (G & S)	1.00	1.20	1.20	1.24	1.27	1.33
Exports (G & S)	1.00	1.17	1.20	1.23	1.30	1.35
Total expenditures	1.00	1.16	1.24	1.27	1.31	1.35
Government consumption	1.00	1.19	1.34	1.40	1.46	1.52
Private consumption	1.00	1.19	1.26	1.29	1.34	1.38
Fixed investment	1.00	1.14	1.21	1.23	1.25	1.29
Change in stocks	1.00	0.64	3.66	3.16	0.47	0.66
Exchange Rate Index	1.00	1.12	1.20	1.28	1.33	1.43

P = Preliminary.

Sources: Bank of Korea, New System of National Account, 1986
Bank of Korea, Preliminary Estimates of 1985 National Accounts, 1986.

Table A1.3: BALANCE OF PAYMENTS
(US$ million at current prices)

	1980	1981	1982	1983	1984	1985
Exports (G & NFS)	21,152.1	25,531.7	26,411.1	28,702.7	32,027.8	31,365.8
Merchandise (FOB)	17,214.0	20,670.8	20,879.2	23,203.9	26,334.6	26,405.3
NFS	3,938.1	4,860.9	5,531.9	5,498.8	5,693.2	4,960.5
Imports (G & NFS)	26,056.1	29,331.4	28,195.0	29,565.3	32,297.9	30,905.5
Merchandise (FOB)	21,598.1	24,299.1	23,478.6	24,967.4	27,370.5	26,435.7
NFS	4,458.0	5,032.3	4,716.4	4,597.9	4,927.4	4,469.8
Resource Balance	-4,904.0	-3,799.7	-1,783.9	-862.6	-270.1	460.3
Net Factor Income	-2,045.7	-2,786.2	-2,992.9	-2,600.2	-2,735.8	-3,058.4
Factor receipts	770.8	1,001.3	1,053.4	1,037.2	1,221.9	1,310.2
Factor payments	2,816.5	3,787.5	4,046.3	3,637.4	3,957.7	4,368.6
(MLT interest paid)	1,594.6	1,995.8	2,397.1	2,257.1	2,555.0	n.a.
Net Current Transfers	1,635.9	1,991.4	2,167.9	1,979.5	1,947.7	1,701.7
Transfer receipts	2,010.3	2,355.1	2,572.2	2,369.8	2,323.5	2,099.3
Transfer payments	374.4	363.7	404.3	390.3	375.8	397.6
Current Account Balance	-5,320.7	-4,646.0	-2,649.7	-1,606.0	-1,372.9	-739.8
Direct investment	96.2	105.4	100.6	101.4	170.7	250.3
Official grant aid	50.0	79.0	52.0	-	-	-
Net MLT Loans (DRS) /a	2,376.3	3,889.0	2,409.1	3,541.5	3,805.9	4,047.0
Disbursements	3,883.3	5,816.2	4,448.1	5,974.0	6,589.5	7,515.0
Repayments	1,507.0	1,927.2	2,038.9	2,432.4	2,783.6	3,468.0
Net Credit from IMF	593.8	627.9	82.2	160.1	319.0	-235.2
Net short-term capital	1,944.5	-82.3	3.6	893.5	-757.9	-538.0
Capital flows NEI	3,983.0	1,090.0	2,159.0	524.0	-113.0	n.a
Errors and omissions	-338.0	-410.0	-1,301.0	-945.0	-884.0	-878.0
Change in Net Reserves	-863.3	-319.6	-92.7	74.1	-739.9	-100.2

/a Includes private nonguaranteed debt.

Sources: Bank of Korea, Economic Statistics Yearbook;
World Bank, World Debt Tables; and
IMF, International Financial Statistics.

Table A1.4: SOCIAL INDICATORS

	KOREA, REPUBLIC OF			REFERENCE GROUPS (WEIGHTED AVERAGES) /a (MOST RECENT ESTIMATE) /b	
	1960/b	1970/b	MOST RECENT ESTIMATE/b	MIDDLE INCOME ASIA & PACIFIC	MIDDLE INCOME LAT. AMERICA & CAR
AREA (THOUSAND SQ. KM)					
TOTAL	98.5	98.5	98.5	.	.
AGRICULTURAL	20.4	23.2	22.3	.	.
GNP PER CAPITA (US$)	..	..	2010.0	1011.1	1875.9
ENERGY CONSUMPTION PER CAPITA (KILOGRAMS OF OIL EQUIVALENT)	143.0	495.0	1104.0	566.8	993.6
POPULATION AND VITAL STATISTICS					
POPULATION,MID-YEAR (THOUSANDS)	25003.0	32241.0	39951.0	.	.
URBAN POPULATION (% OF TOTAL)	27.7	41.1	62.5	35.9	67.7
POPULATION PROJECTIONS					
POPULATION IN YEAR 2000 (MILL)			50.5	.	.
STATIONARY POPULATION (MILL)			70.0	.	.
POPULATION MOMENTUM			1.6	.	.
POPULATION DENSITY					
PER SQ. KM.	253.9	327.4	405.7	386.9	48.0
PER SQ. KM. AGRI. LAND	1223.2	1387.3	1763.9	1591.2	91.1
POPULATION AGE STRUCTURE (%)					
0-14 YRS	42.9	42.0	31.5	38.2	38.5
15-64 YRS	53.7	54.6	64.2	57.7	57.1
65 AND ABOVE	3.3	3.2	4.2	3.5	4.2
POPULATION GROWTH RATE (%)					
TOTAL	2.1	2.5	1.6	2.3	2.4
URBAN	4.7	6.4	4.4	4.1	3.6
CRUDE BIRTH RATE (PER THOUS)	42.8	30.4	22.8	30.1	30.9
CRUDE DEATH RATE (PER THOUS)	13.7	9.6	6.2	9.4	8.0
GROSS REPRODUCTION RATE	2.8	2.1	1.3	1.9	2.0
FAMILY PLANNING					
ACCEPTORS, ANNUAL (THOUS)	..	671.0	..	.	.
USERS (% OF MARRIED WOMEN)	..	25.0	58.0	56.5	45.3
FOOD AND NUTRITION					
INDEX OF FOOD PROD. PER CAPITA (1969-71=100)	89.0	99.0	125.0	124.4	109.6
PER CAPITA SUPPLY OF					
CALORIES (% OF REQUIREMENTS)	88.0	103.0	126.0	115.7	113.2
PROTEINS (GRAMS PER DAY)	53.0	62.0	83.0	60.3	69.4
OF WHICH ANIMAL AND PULSE	7.0	8.0	14.0 /c	14.1	34.2
CHILD (AGES 1-4) DEATH RATE	7.1	4.9	2.0	7.2	4.8
HEALTH					
LIFE EXPECT. AT BIRTH (YEARS)	53.9	59.1	67.4	60.6	64.8
INFANT MORT. RATE (PER THOUS)	78.5	50.0	29.0	64.9	59.7
ACCESS TO SAFE WATER (%POP)					
TOTAL	12.1	58.0	78.0 /d	46.0	65.3
URBAN	18.6	84.0	86.0 /d	57.6	76.5
RURAL	9.5	38.0	61.0 /d	37.1	44.2
ACCESS TO EXCRETA DISPOSAL (% OF POPULATION)					
TOTAL	..	25.0	100.0 /d	50.1	56.3
URBAN	..	59.0	100.0 /d	52.9	73.4
RURAL	..	..	100.0 /d	44.7	25.5
POPULATION PER PHYSICIAN	3540.0	2240.0	1440.0	7751.7	1909.7
POP. PER NURSING PERSON	3240.0 /e,f	1790.0 /f	350.0	2464.8	808.2
POP. PER HOSPITAL BED					
TOTAL	2510.0	1950.0	960.0	1112.1	362.0
URBAN	1280.0 /e	1100.0	750.0 /g	651.4	422.0
RURAL	..	..	..	2596.9	2716.7
ADMISSIONS PER HOSPITAL BED	..	14.9	..	41.1	27.5
HOUSING					
AVERAGE SIZE OF HOUSEHOLD					
TOTAL	5.6	5.2	4.5	..	..
URBAN	5.4	4.9	4.4	..	..
RURAL	5.6	5.5	4.7	..	..
AVERAGE NO. OF PERSONS/ROOM					
TOTAL	2.5	2.3	..	..	..
URBAN	2.8	2.7	..	..	..
RURAL	2.4	2.2	..	..	..
PERCENTAGE OF DWELLINGS WITH ELECT.					
TOTAL	28.4	49.9	..	..	..
URBAN	67.3	92.4	..	..	..
RURAL	12.4	29.9	64.9 /c	..	..

Table A1.4: SOCIAL INDICATORS

	KOREA, REPUBLIC OF			REFERENCE GROUPS (WEIGHTED AVERAGES) /a (MOST RECENT ESTIMATE) /b	
	1960/b	1970/b	MOST RECENT ESTIMATE/b	MIDDLE INCOME ASIA & PACIFIC	MIDDLE INCOME LAT. AMERICA & CAR
EDUCATION					
ADJUSTED ENROLLMENT RATIOS					
PRIMARY: TOTAL	94.0	103.0	100.0	100.7	106.7
MALE	99.0	104.0	102.0	104.4	108.5
FEMALE	89.0	103.0	99.0	97.2	104.6
SECONDARY: TOTAL	27.0	42.0	89.0	47.8	44.2
MALE	38.0	50.0	94.0	50.6	42.7
FEMALE	14.0	32.0	85.0	44.8	44.9
VOCATIONAL (% OF SECONDARY)	14.2	14.3	18.1	18.4	13.3
PUPIL-TEACHER RATIO					
PRIMARY	58.0	57.0	42.0	30.4	29.9
SECONDARY	34.0	37.0	29.0	22.2	16.7
CONSUMPTION					
PASSENGER CARS/THOUSAND POP	0.5	1.9	6.5 /d	10.1	46.0
RADIO RECEIVERS/THOUSAND POP	31.2	124.4	432.2	172.9	328.3
TV RECEIVERS/THOUSAND POP	0.3	13.0	174.1	58.5	112.4
NEWSPAPER ("DAILY GENERAL INTEREST") CIRCULATION PER THOUSAND POPULATION	68.2	136.3	192.3	65.3	81.1
CINEMA ANNUAL ATTENDANCE/CAPITA	4.1	5.2	1.1	3.4	2.4
LABOR FORCE					
TOTAL LABOR FORCE (THOUS)	8304.0	11285.0	16018.0	.	.
FEMALE (PERCENT)	26.1	32.7	32.5	33.6	23.6
AGRICULTURE (PERCENT)	66.0	50.0	34.0 /d	52.2	31.4
INDUSTRY (PERCENT)	9.0	17.0	29.0 /d	17.9	24.3
PARTICIPATION RATE (PERCENT)					
TOTAL	33.2	35.0	40.1	38.9	33.5
MALE	49.5	46.8	53.1	50.8	51.3
FEMALE	17.2	23.0	26.1	26.8	15.9
ECONOMIC DEPENDENCY RATIO	1.4	1.3	0.9	1.1	1.3
INCOME DISTRIBUTION					
PERCENT OF PRIVATE INCOME RECEIVED BY					
HIGHEST 5% OF HOUSEHOLDS	..	17.1	16.1 /g	..	..
HIGHEST 20% OF HOUSEHOLDS	..	44.5	45.3 /g	48.0	..
LOWEST 20% OF HOUSEHOLDS	..	7.1	5.7 /g	6.4	..
LOWEST 40% OF HOUSEHOLDS	..	17.7	16.9 /g	15.5	..
POVERTY TARGET GROUPS					
ESTIMATED ABSOLUTE POVERTY INCOME LEVEL (US$ PER CAPITA)					
URBAN	..	..	320.0 /h	..	288.3
RURAL	..	..	270.0 /h	151.9	185.3
ESTIMATED RELATIVE POVERTY INCOME LEVEL (US$ PER CAPITA)					
URBAN	..	..	370.0 /h	177.9	519.8
RURAL	..	..	310.0 /h	164.7	359.7
ESTIMATED POP. BELOW ABSOLUTE POVERTY INCOME LEVEL (%)					
URBAN	..	..	18.0 /h	23.5	..
RURAL	..	..	11.0 /h	37.8	..

.. NOT AVAILABLE
. NOT APPLICABLE

NOTES

/a The group averages for each indicator are population-weighted arithmetic means. Coverage of countries among the indicators depends on availability of data and is not uniform.

/b Unless otherwise noted, "Data for 1960" refer to any year between 1959 and 1961; "Data for 1970" between 1969 and 1971; and data for "Most Recent Estimate" between 1981 and 1983.

/c 1977; /d 1980; /e 1962; /f Registered, not all practising in the country; /g 1976; /h 1978.

JUNE, 1985

DEFINITIONS OF SOCIAL INDICATORS

Notes: Although the data are drawn from sources generally judged the most authoritative and reliable, it should also be noted that they may not be internationally comparable because of the lack of standardized definitions and concepts used by different countries in collecting the data. The data are, nonetheless, useful to describe orders of magnitude, indicate trends, and characterize certain major differences between countries.

The reference groups are (1) the same country group of the subject country and (2) a country group with somewhat higher average income than the country group of the subject country (except for "High Income Oil Exporters" group where "Middle Income North Africa and Middle East" is chosen because of stronger socio-cultural affinities). In the reference group data the averages are population weighted arithmetic means for each indicator and shown only when majority of the countries in a group has data for that indicator. Since the coverage of countries among the indicators depends on the availability of data and is not uniform, caution must be exercised in relating averages of one indicator to another. These averages are only useful in comparing the value of one indicator at a time among the country and reference groups.

AREA (thousand sq.km.)

Total—Total surface area comprising land area and inland waters; 1960, 1970 and 1983 data.

Agricultural—Estimate of agricultural area used temporarily or permanently for crops, pastures, market and kitchen gardens or to lie fallow, 1960, 1970 and 1982 data.

GNP PER CAPITA (US$)—GNP per capita estimates at current market prices, calculated by same conversion method as *World Bank Atlas* (1981-83 basis); 1983 data.

ENERGY CONSUMPTION PER CAPITA—Annual apparent consumption of commercial primary energy (coal and lignite, petroleum, natural gas and hydro-, nuclear and geothermal electricity) in kilograms of oil equivalent per capita; 1960, 1970, and 1982 data.

POPULATION AND VITAL STATISTICS

Total Population, Mid-Year (thousands)—As of July 1; 1960, 1970, and 1983 data.

Urban Population (percent of total)—Ratio of urban to total population; different definitions of urban areas may affect comparability of data among countries; 1960, 1970, and 1983 data.

Population Projections

Population in year 2000—The projection of population for 2000, made for each economy separately. Starting with information on total population by age and sex, fertility rates, mortality rates, and international migration in the base year 1980, these parameters were projected at five-year intervals on the basis of generalized assumptions until the population became stationary.

Stationary population—Is one in which age- and sex-specific mortality rates have not changed over a long period, while age-specific fertility rates have simultaneously remained at replacement level (net reproduction rate = 1). In such a population, the birth rate is constant and equal to the death rate, the age structure is also constant, and the growth rate is zero. The stationary population size was estimated on the basis of the projected characteristics of the population in the year 2000, and the rate of decline of fertility rate to replacement level.

Population Momentum—Is the tendency for population growth to continue beyond the time that replacement-level fertility has been achieved; that is, even after the net reproduction rate has reached unity. The momentum of a population in the year *t* is measured as a ratio of the ultimate stationary population to the population in the year *t*, given the assumption that fertility remains at replacement level from year *t* onward, 1985 data.

Population Density

Per sq.km.—Mid-year population per square kilometer (100 hectares) of total area; 1960, 1970, and 1983 data.

Per sq.km. agricultural land—Computed as above for agricultural land only, 1960, 1970, and 1982 data.

Population Age Structure (percent)—Children (0-14 years), working age (15-64 years), and retired (65 years and over) as percentage of mid-year population; 1960, 1970, and 1983 data.

Population Growth Rate (percent)—total—Annual growth rates of total mid-year population for 1950-60, 1960-70, and 1970-83.

Population Growth Rate (percent)—urban—Annual growth rates of urban population for 1950-60, 1960-70, and 1970-83 data.

Crude Birth Rate (per thousand)—Number of live births in the year per thousand of mid-year population; 1960, 1970, and 1983 data.

Crude Death Rate (per thousand)—Number of deaths in the year per thousand of mid-year population; 1960, 1970, and 1983 data.

Gross Reproduction Rate—Average number of daughters a woman will bear in her normal reproductive period if she experiences present age-specific fertility rates; usually five-year averages ending in 1960, 1970, and 1983.

Family Planning—Acceptors, Annual (thousands)—Annual number of acceptors of birth-control devices under auspices of national family planning program.

Family Planning—Users (percent of married women)—The percentage of married women of child-bearing age who are practicing or whose husbands are practicing any form of contraception. Women of child-bearing age are generally women aged 15-49, although for some countries contraceptive usage is measured for other age groups.

FOOD AND NUTRITION

Index of Food Production Per Capita (1969-71 = 100)—Index of per capita annual production of all food commodities. Production excludes animal feed and seed for agriculture. Food commodities include primary commodities (e.g. sugarcane instead of sugar) which are edible and contain nutrients (e.g. coffee and tea are excluded); they comprise cereals, root crops, pulses, oil seeds, vegetables, fruits, nuts, sugarcane and sugar beets, livestock, and livestock products. Aggregate production of each country is based on national average producer price weights; 1961-65, 1970, and 1982 data.

Per Capita Supply of Calories (percent of requirements)—Computed from calorie equivalent of net food supplies available in country per capita per day. Available supplies comprise domestic production, imports less exports, and changes in stock. Net supplies exclude animal feed, seeds for use in agriculture, quantities used in food processing, and losses in distribution. Requirements were estimated by FAO based on physiological needs for normal activity and health considering environmental temperature, body weights, age and sex distribution of population, and allowing 10 percent for waste at household level; 1961, 1970 and 1982 data.

Per Capita Supply of Protein (grams per day)—Protein content of per capita net supply of food per day. Net supply of food is defined as above. Requirements for all countries established by USDA provide for minimum allowances of 60 grams of total protein per day and 20 grams of animal and pulse protein, of which 10 grams should be animal protein. These standards are lower than those of 75 grams of total protein and 23 grams of animal protein as an average for the world, proposed by FAO in the Third World Food Supply; 1961, 1970 and 1982 data.

Per Capita Protein Supply From Animal and Pulse—Protein supply of food derived from animals and pulses in grams per day; 1961-65, 1970 and 1977 data.

Child (ages 1-4) Death Rate (per thousand)—Number of deaths of children aged 1-4 years per thousand children in the same age group in a given year. For most developing countries data derived from life tables; 1960, 1970 and 1983 data.

HEALTH

Life Expectancy at Birth (years)—Number of years a newborn infant would live if prevailing patterns of mortality for all people at the time of of its birth were to stay the same throughout its life; 1960, 1970 and 1983 data.

Infant Mortality Rate (per thousand) - Number of infants who die before reaching one year of age per thousand live births in a given year; 1960, 1970 and 1983 data.

Access to Safe Water (percent of population)—total, urban, and rural - Number of people (total, urban, and rural) with reasonable access to safe water supply (includes treated surface waters or untreated but uncontaminated water such as that from protected boreholes, springs and sanitary wells) as percentages of their respective populations. In an urban area a public fountain or standpost located not more than 200 meters from a house may be considered as being within reasonable access of that house. In rural areas reasonable access would imply that the housewife or members of the houselhold do not have to spend a disproportionate part of the day in fetching the family's water needs.

Access to Excreta Disposal (percent of population)—total, urban, and rural - Number of people (total, urban, and rural) served by excreta disposal as percentages of their respective populations. Excreta disposal may include the collection and disposal, with or without treatment, of human excreta and waste-water by water-borne systems or the use of pit privies and similar installations.

Population per Physician—Population divided by number of practising physicians qualified from a medical school at university level.

Population per Nursing Person—Population divided by number of practicing male and female graduate nurses, assistant nurses, practical nurses and nursing auxiliaries.

Population per Hospital Bed—total, urban, and rural—Population (total, urban, and rural) divided by their respective number of hospital beds available in public and private, general and specialized hospitals and rehabilitation centers. Hospitals are establishments permanently staffed by at least one physician. Establishments providing principally custodial care are not included. Rural hospitals, however, include health and medical centers not permanently staffed by a physician (but by a medical assistant, nurse, midwife, etc.) which offer in-patient accommodation and provide a limited range of medical facilities.

Admissions per Hospital Bed—Total number of admissions to or discharges from hospitals divided by the number of beds.

HOUSING

Average Size of Household (persons per household)—total, urban, and rural—A household consists of a group of individuals who share living quarters and their main meals. A boarder or lodger may or may not be included in the household for statistical purposes.

Average Number of Persons per Room—total, urban, and rural—Average number of persons per room in all urban, and rural occupied conventional dwellings, respectively. Dwellings exclude non-permanent structures and unoccupied parts.

Percentage of Dwellings with Electricity—total, urban, and rural—Conventional dwellings with electricity in living quarters as percentage of total, urban, and rural dwellings respectively.

EDUCATION

Adjusted Enrollment Ratios

Primary school - total, male and female—Gross total, male and female enrollment of all ages at the primary level as percentages of respective primary school-age populations. While many countries consider primary school age to be 6-11 years, others do not. The differences in country practices in the ages and duration of school are reflected in the ratios given. For some countries with universal education, gross enrollment may exceed 100 percent since some pupils are below or above the country's standard primary-school age.

Secondary school - total, male and female—Computed as above; secondary education requires at least four years of approved primary instruction; provides general, vocational, or teacher training instructions for pupils usually of 12 to 17 years of age; correspondence courses are generally excluded.

Vocational Enrollment (percent of secondary)—Vocational institutions include technical, industrial, or other programs which operate independently or as departments of secondary institutions.

Pupil-teacher Ratio - primary, and secondary - Total students enrolled in primary and secondary levels divided by numbers of teachers in the corresponding levels.

CONSUMPTION

Passenger Cars (per thousand population) - Passenger cars comprise motor cars seating less than eight persons; excludes ambulances, hearses and military vehicles.

Radio Receivers (per thousand population) --All types of receivers for radio broadcasts to general public per thousand of population; excludes un-licensed receivers in countries and in years when registration of radio sets was in effect; data for recent years may not be comparable since most countries abolished licensing.

TV Receivers (per thousand population) - TV receivers for broadcast to general public per thousand population; excludes unlicensed TV receivers in countries and in years when registration of TV sets was in effect.

Newspaper Circulation (per thousand population)—Shows the average circulation of "daily general interest newspaper," defined as a periodical publication devoted primarily to recording general news. It is considered to be "daily" if it appears at least four times a week.

Cinema Annual Attendance per Capita per Year Based on the number of tickets sold during the year, including admissions to drive-in cinemas and mobile units.

LABOR FORCE

Total Labor Force (thousands)—Economically active persons, including armed forces and unemployed but excluding housewives, students, etc., covering population of all ages. Definitions in various countries are not comparable; 1960, 1970 and 1983 data.

Female (percent)—Female labor force as percentage of total labor force.

Agriculture (percent)—Labor force in farming, forestry, hunting and fishing as percentage of total labor force; 1960, 1970 and 1980 data.

Industry (percent)—Labor force in mining, construction, manufacturing and electricity, water and gas as percentage of total labor force; 1960, 1970 and 1980 data.

Participation Rate (percent)—total, male, and female—Participation or activity rates are computed as total, male, and female labor force as percentages of total, male and female population of all ages respectively; 1960, 1970, and 1983 data. These are based on ILO's participation rates reflecting age-sex structure of the population, and long time trend. A few estimates are from national sources.

Economic Dependency Ratio—Ratio of population under 15, and 65 and over, to the working age population (those aged 15-64).

INCOME DISTRIBUTION

Percentage of Total Disposable Income (both in cash and kind)—Accruing to percentile groups of households ranked by total household income.

POVERTY TARGET GROUPS

The following estimates are very approximate measures of poverty levels, and should be interpreted with considerable caution.

Estimated Absolute Poverty Income Level (US$ per capita)—urban and rural—Absolute poverty income level is that income level below which a minimal nutritionally adequate diet plus essential non-food requirements is not affordable.

Estimated Relative Poverty Income Level (US$ per capita)—urban and rural—Rural relative poverty income level is one-third of average per capita personal income of the country. Urban level is derived from the rural level with adjustment for higher cost of living in urban areas.

Estimated Population Below Absolute Poverty Income Level (percent)—urban and rural— Percent of population (urban and rural who are "absolute poor."

Comparative Analysis and Data Division
Economic Analysis and Projections Department
June 1985

Table A2.1: POPULATION AND EMPLOYMENT

Year	Total population (mid-year)	Pop. 14y+	Total labor force ('000)	Employed ('000)
1965	28,705	15,937	8,859	8,206
1970	32,241	18,253	10,199	9,745
1975	35,281	21,835	12,340	11,830
1976	35,849	22,549	13,061	12,556
1977	36,412	23,336	13,440	12,929
1978	36,969	24,024	13,932	13,490
1979	37,534	24,678	14,206	13,664
1980	38,124	25,335	14,454	13,706
1981	38,723	25,969	14,710	14,048
1982	39,331	26,531	15,080	14,424
1983	39,951	27,130	15,128	14,515
1984	40,578	27,793	14,984	14,417
1985	41,056	28,489	15,554	14,935
Growth rate p.a. (%)				
1965-70	2.4	3.0	2.9	3.5
1970-75	1.8	3.2	3.9	4.0
1975-80	1.6	2.3	4.5	4.5
1981	1.6	2.5	1.8	2.5
1982	1.6	2.2	2.5	2.7
1983	1.6	2.3	0.3	0.6
1984	1.6	2.4	-1.0	-0.7
1985	1.2	2.5	3.8	3.6

Year	Labor force participation rate: Average (%)	Male (%)	Female (%)	Unemployment rate (%)
1965	55.6	76.6	36.5	7.4
1970	55.5	75.1	38.5	4.5
1975	56.5	74.5	36.9	4.1
1976	57.9	74.6	42.3	3.9
1977	57.6	75.9	40.7	3.8
1978	58.0	75.0	42.2	3.2
1979	57.6	74.1	42.2	3.8
1980	57.1	73.6	41.6	5.2
1981	56.6	73.2	41.1	4.5
1982	56.8	72.5	42.2	4.4
1983	55.8	70.9	41.6	4.1
1984	53.9	69.4	39.5	3.8
1985p	54.2	69.6	40.7	4.0

Source: Economic Planning Board, Korea Statistical Yearbook.

Table A2.2: SECTORAL EMPLOYMENT

Year	Total employment ('000)	Agriculture		Mining and manufacturing		Other	
		'000	%	'000	%	'000	%
1965	8,206	4,810	58.6	849	10.3	2,547	31.0
1970	9,745	4,916	50.4	1,395	14.4	3,434	35.2
1975	11,830	5,425	45.9	2,265	19.1	4,140	35.0
1976	12,556	5,601	44.6	2,743	21.9	4,212	33.5
1977	12,929	5,405	41.8	2,901	22.4	4,623	35.8
1978	13,490	5,181	38.4	3,123	23.2	5,186	38.4
1979	13,664	4,887	35.8	3,237	23.7	5,540	40.5
1980	13,706	4,658	34.0	3,095	22.6	5,952	43.4
1981	14,048	4,806	34.2	2,996	21.3	6,247	44.5
1982	14,424	4,623	32.1	3,157	21.9	6,644	46.1
1983	14,515	4,314	29.7	3,383	23.3	6,818	47.0
1984	14,417	3,909	27.1	3,493	24.2	7,015	48.7
1985	14,935	3,722	24.9	3,654	24.5	7,559	50.6
Growth rate p.a. (%)							
1965-70	3.5	0.4		10.4		6.2	
1970-75	4.0	2.0		10.2		3.8	
1975-80	3.0	-3.0		6.4		7.5	
1976	6.1	3.2		21.1		1.7	
1977	3.0	-3.5		5.8		9.8	
1978	4.3	-4.1		7.7		12.2	
1979	1.3	-5.7		3.7		6.8	
1980	0.3	-4.7		-4.4		7.4	
1981	2.5	3.2		-3.2		5.0	
1982	2.7	9.6		5.4		6.4	
1983	0.6	-6.7		7.2		2.6	
1984	-0.7	-9.4		3.3		2.9	
1985	3.6	-4.8		4.6		7.8	

Source: Economic Planning Board, Korea Statistical Yearbook.

Table A3.1: GNP BY EXPENDITURE CATEGORY (CURRENT PRICES), 1980-85
(billion won)

	1980	1981	1982	1983	1984	1985P
Current Prices						
Consumption	29,054.1	35,881.0	40,111.6	44,035.3	47,857.4	51,787.4
Private consumption	24,786.1	30,497.6	34,001.3	37,281.9	40,778.3	44,005.2
Public consumption	4,268.0	5,383.4	6,110.3	6,753.4	7,079.1	7,782.2
Gross Domestic Investment	11,788.9	13,679.0	14,509.6	17,620.8	21,207.3	22,540.0
Fixed investment	11,835.7	12,931.0	15,486.5	18,479.6	20,795.0	22,091.0
Change in stocks	-46.8	748.0	-976.9	-858.8	412.3	449.0
Exports of goods and services	12,765.4	17,191.5	18,802.3	22,245.7	25,829.6	27,328.5
Imports of goods and services	15,729.3	19,712.3	20,153.6	23,027.9	26,037.2	26,904.1
Statistical discrepancy	35.8	-15.5	-357.2	129.0	9.6	226.1
Gross Domestic Product	37,914.9	47,023.7	52,912.7	61,002.9	68,866.7	74,977.9
Net factor income	-1,242.6	-1,897.5	-2,188.1	-2,017.1	-2,458.5	-2,660.9
Gross National Product	36,672.3	45,126.2	50,724.6	58,985.8	66,408.2	72,317.0
Composition (% at Current Prices)						
Consumption	79.2	79.5	79.1	74.7	72.1	71.6
Private consumption	67.6	67.6	67.0	63.2	61.4	60.9
Public consumption	11.6	11.9	12.1	11.5	10.7	10.7
Gross Domestic Investment	32.1	30.3	28.6	29.9	31.9	31.2
Fixed investment	32.3	28.7	30.5	31.3	31.3	30.5
Change in stocks	-0.2	1.6	-1.9	-1.4	0.6	0.7
Exports of goods and services	34.8	38.1	37.1	37.7	38.9	37.8
Imports of goods and services	42.9	43.7	39.7	39.0	39.2	37.2
Statistical discrepancy	0.1	0.0	-0.7	0.2	0.0	0.3
Gross Domestic Product	103.4	104.2	104.3	103.4	103.7	103.7
Net factor income	-3.4	-4.2	-4.3	-3.4	-3.7	-3.7
Gross National Product	100.0	100.0	100.0	100.0	100.0	100.0

P = Preliminary.

Source: Bank of Korea, New System of National Accounts, 1986.
Bank of Korea, Preliminary Estimates of 1985 National Accounts, 1986.

Table A3.2: GNP BY EXPENDITURE CATEGORY (1980 CONSTANT PRICES), 1980-85
(billion won)

	1980	1981	1982	1983	1984	1985P
Constant Prices						
Consumption	29,054.1	30,110.2	31,367.1	33,626.7	35,373.3	37,086.1
Private consumption	24,786.1	25,578.3	26,798.0	28,815.8	30,538.3	31,956.6
Public consumption	4,268.0	4,531.9	4,569.1	4,810.9	4,835.0	5,129.5
Gross Domestic Investment	11,788.9	12,535.9	12,553.2	14,745.0	17,492.1	17,768.9
Fixed investment	11,835.7	11,359.5	12,820.1	15,016.9	16,617.5	17,085.0
Change in stocks	-46.8	1,176.4	-266.9	-271.9	874.6	683.9
Exports of goods and services	12,765.4	14,684.5	15,637.9	18,054.0	19,854.7	20,316.1
Imports of goods and services	15,729.3	16,432.1	16,763.3	18,593.4	20,464.7	20,155.8
Statistical discrepancy	35.8	-175.2	240.8	-88.6	-382.9	-470.5
Gross Domestic Product	37,914.9	40,723.3	43,035.7	47,743.7	51,872.5	54,544.8
Net factor income	-1,242.6	-1,634.6	-1,824.1	-1,634.6	-1,869.5	-1,971.7
Gross National Product	36,672.3	39,088.7	41,211.6	46,109.1	50,003.0	52,573.1
Growth Rate (% p.a. at Constant Prices)						
Consumption		3.6	4.2	7.2	5.2	4.8
Private consumption		3.2	4.8	7.5	6.0	4.6
Public consumption		6.2	0.8	5.3	0.5	6.1
Gross Domestic Investment		6.3	0.1	17.5	10.1	1.6
Fixed investment		-4.0	12.9	17.1	10.7	2.8
Change in stocks		2,613.7	-122.7	-1.9	421.7	-21.8
Exports of goods and services		15.0	6.5	15.5	10.0	2.3
Imports of goods and services		4.5	2.0	10.9	10.1	-1.5
Statistical discrepancy		-	-	-	-	-
Gross Domestic Product		7.4	5.7	10.9	8.6	5.2
Net factor income		-	-	-	-	-
Gross National Product		6.6	5.4	11.9	8.4	5.1

P = Preliminary.

Source: Bank of Korea, New System of National Accounts, 1986.
Bank of Korea, Preliminary Estimates of 1985 National Accounts, 1986.

Table A3.3: GNP (AT MARKET PRICES) BY INDUSTRIAL ORIGIN, 1980-85
(Billion won at current prices)

	1980	1981	1982	1983	1984	1985P
Current Prices						
Industry	33,250.7	41,134.5	45,964.3	52,835.0	60,059.4	65,441.1
Agriculture, forestry & fishing	5,524.7	7,442.3	7,731.5	8,292.6	9,181.2	10,353.4
Mining & quarrying	520.0	734.3	772.6	873.1	977.8	1,088.9
Manufacturing	11,214.3	13,713.8	14,996.2	17,301.7	20,019.4	21,070.9
Electricity, gas & water	786.2	995.1	1,159.0	1,514.0	1,937.9	2,310.7
Construction	3,149.8	3,443.3	4,221.5	5,075.4	5,683.0	6,087.9
Wholesale & retail trade, restaurants & hotels	4,967.5	6,166.9	6,096.8	7,743.7	8,872.1	9,671.5
Transport, storage & communications	2,915.4	3,893.7	4,616.5	5,277.4	5,789.6	6,193.6
Financing, insurance, real estate & business services	4,171.7	4,480.4	4,462.3	5,873.1	6,701.6	7,878.1
Social & personal services	1,192.5	1,442.0	1,760.5	2,112.9	2,406.6	2,656.7
Less: Imputed financial services	1,191.4	1,177.3	662.6	1,228.9	1,509.8	1,867.6
Government services	2,823.6	3,522.2	4,226.5	4,711.1	5,060.2	5,539.2
Private nonprofit institution services	705.3	920.1	1,168.1	1,391.1	1,566.9	1,773.1
Rest of the world	-1,242.6	-1,897.5	-2,188.1	-2,017.1	-2,458.5	-2,660.0
Gross National Product	36,672.3	45,126.2	50,724.6	58,985.8	66,408.2	72,317.0
Composition (% at Current Prices)						
Industry	90.7	91.2	90.6	89.6	90.4	90.5
Agriculture, forestry & fishing	15.1	16.5	15.2	14.1	13.8	14.3
Mining & quarrying	1.4	1.6	1.5	1.5	1.5	1.5
Manufacturing	30.6	30.4	29.6	29.3	30.1	29.1
Electricity, gas & water	2.1	2.2	2.3	2.6	2.9	3.2
Construction	8.6	7.6	8.3	8.6	8.6	8.4
Wholesale & retail trade, restaurants & hotels	13.5	13.7	13.6	13.1	13.4	13.4
Transport, storage & communications	7.9	8.6	9.1	8.9	8.7	8.6
Financing, insurance, real estate & business services	11.4	9.9	8.8	10.0	10.1	10.9
Social and personal services	3.3	3.2	3.5	3.6	3.6	3.7
Less: Imputed financial services	3.2	2.6	1.3	2.1	2.3	2.6
Government services	7.7	7.8	8.3	8.0	7.6	7.7
Private nonprofit institution services	1.9	2.0	2.3	2.4	2.4	2.5
Rest of the world	-3.4	-4.2	-4.3	-3.4	-3.7	-3.7
Gross National Product	100.0	100.0	100.0	100.0	100.0	100.0

P = Preliminary

Source: Bank of Korea, New System of National Accounts, 1986.
Bank of Korea, Preliminary Estimates of 1985 National Accounts, 1986.

Table A3.4: **GNP (AT MARKET PRICES) BY INDUSTRIAL ORIGIN, 1980-85**
(Billion won at constant prices)

	1980	1981	1982	1983	1984	1985P
Current Prices						
Industry	33,250.7	35,801.6	37,891.6	42,239.1	46,238.6	48,798.8
Agriculture, forestry & fishing	5,524.7	6,759.7	6,980.6	7,436.0	7,453.2	7,893.4
Mining & quarrying	520.0	616.1	626.4	685.0	742.4	794.0
Manufacturing	11,214.3	12,058.9	12,558.9	14,095.6	16,188.0	16,757.8
Electricity, gas & water	786.2	844.4	881.5	1,131.1	1,420.3	1,740.1
Construction	3,149.8	3,035.3	3,594.9	4,275.6	4,539.5	4,632.7
Wholesale & retail trade, restaurants & hotels	4,967.5	5,353.7	5,637.0	6,141.9	6,716.0	7,073.7
Transport, storage & communications	2,915.4	3,072.3	3,288.1	3,668.4	3,974.2	4,164.9
Financing, insurance, real estate & business services	4,171.7	4,128.0	4,464.3	4,945.8	5,468.5	6,197.8
Social and personal services	1,192.5	1,210.4	1,304.2	1,452.8	1,590.1	1,695.4
Less: Imputed financial services	1,191.4	1,277.2	1,444.3	1,593.1	1,853.6	2,151.0
Government services	2,823.6	2,934.1	3,022.6	3,090.8	3,112.4	3,186.0
Private nonprofit institution services	705.3	739.3	793.0	857.7	925.8	982.2
Rest of the world	-1,242.6	-1,634.6	-1,824.1	-1,634.6	-1,869.5	-1,971.7
Gross National Product	36,672.3	39,088.7	41,211.6	46,109.1	50,003.0	52,573.1
Growth Rate (% p.a. at Constant Prices)						
Industry		7.7	5.8	11.5	9.5	5.5
Agriculture, forestry & fishing		24.4	3.3	6.5	0.2	5.9
Mining & quarrying		18.5	1.7	9.4	8.4	7.0
Manufacturing		7.5	4.1	12.2	14.3	3.5
Electricity, gas & water		7.4	4.4	28.3	25.6	22.5
Construction		-3.6	18.4	18.9	6.2	2.1
Wholesale & retail trade, restaurants & hotels		7.8	5.3	9.0	9.3	5.3
Transport, storage & communications		5.4	7.0	11.6	8.3	4.8
Financing, insurance, real estate & business services		-1.0	2.1	10.8	10.6	13.3
Social and personal services		1.5	7.7	11.4	9.5	6.6
Less: Imputed financial services		7.2	13.1	10.3	16.4	16.0
Government services		3.9	3.0	2.3	0.7	2.4
Private nonprofit institution services		4.8	7.3	8.2	7.9	6.1
Rest of the world		-	-	-	-	-
Gross National Product		6.6	5.4	11.9	8.4	5.1

P = Preliminary

Source: Bank of Korea, **New System of National Accounts**, 1986.
Bank of Korea, **Preliminary Estimates of 1985 National Accounts**, 1986.

Table A3.5: GDP (AT FACTOR COST) BY INDUSTRIAL ORIGIN (CURRENT PRICES), 1980-84
(Billion won)

	1980	1981	1982	1983	1984
Current Prices					
Industry	27,085.2	33,477.2	36,718.0	41,801.3	47,695.9
Agriculture, forestry & fishing	5,342.0	7,204.3	7,458.1	7,866.9	8,724.6
Mining & quarrying	526.8	683.6	656.2	711.2	790.4
Manufacturing	7,561.7	9,253.7	9,977.9	11,524.8	13,460.7
Electricity, gas & water	589.3	753.3	790.4	1,053.7	1,411.5
Construction	2,750.0	2,951.5	3,631.5	4,364.6	4,914.3
Wholesale & retail trade, restaurants & hotels	4,758.2	5,900.9	6,538.2	7,390.7	8,549.2
Transport, storage & communications	2,243.1	3,029.3	3,474.3	3,953.5	4,254.0
Financing, insurance, real estate & business services	3,473.9	3,637.2	3,348.1	4,358.6	5,033.4
Social & personal services	1,031.6	1,240.7	1,505.9	1,806.2	2,067.6
Less: Imputed financial services	-1,191.4	-1,177.3	-662.6	-1,228.9	-1,509.8
Government services	2,735.7	3,413.0	4,093.9	4,573.4	4,914.4
Private nonprofit institution services	659.2	858.3	1,091.6	1,301.1	1,466.6
Gross Domestic Product at F.C.	30,480.1	37,748.5	41,903.5	47,675.8	54,076.9
Composition (% at Current Prices)					
Industry	88.9	88.7	87.6	87.7	88.2
Agriculture, forestry & fishing	17.5	19.1	17.8	16.5	16.1
Mining & quarrying	1.7	1.8	1.6	1.5	1.5
Manufacturing	24.8	24.5	23.8	24.2	24.9
Electricity, gas & water	2.2	2.0	1.9	2.2	2.6
Construction	9.0	7.8	8.7	9.2	9.1
Wholesale & retail trade, restaurants & hotels	15.6	15.6	15.6	15.5	15.8
Transport, storage & communications	8.3	8.0	8.3	8.3	7.9
Financing, insurance, real estate & business services	11.4	9.6	8.0	9.1	9.3
Social & personal services	3.4	3.3	3.6	3.8	3.8
Less: Imputed financial services	-3.9	-3.1	-1.6	-2.6	-2.8
Government services	9.0	9.0	9.8	9.6	9.1
Private nonprofit institution services	2.2	2.3	2.6	2.7	2.7
Gross National Product	100.0	100.0	100.0	100.0	100.0

Source: Bank of Korea, New System of National Accounts, 1986.
Bank of Korea, Preliminary Estimates of 1985 National Accounts, 1986.

Table A3.6: GDP (AT FACTOR COST) BY INDUSTRIAL ORIGIN (1980 DATA BASE) CONSTANT PRICES, 1970-84 /a

	1970	1975	1980	1981	1982	1983	1984
Constant Prices							
Agriculture, forestry & fishing	4,944.51	6,276.27	5,344.15	6,653.47	6,926.26	7,362.61	7,384.70
Fishing	286.75	460.90	512.49	590.90	600.95		
Mining & quarrying	407.87	576.32	581.93	675.62	649.95	709.75	769.37
Manufacturing	1,939.30	4,444.38	8,452.28	9,060.84	9,423.27	10,450.41	11,976.17
Electricity, gas & water	156.87	345.61	700.96	768.95	833.54	942.73	1,039.83
Electricity & gas	144.93	322.75	660.12	722.17	783.55		
Construction	1,152.80	1,716.32	2,994.25	2,832.56	3,399.07	4,119.67	4,115.55
Wholesale & retail trade, restaurants & hotels	2,175.09	3,775.43	5,146.80	5,476.20	5,722.63	6,363.56	7,038.10
Transport, storage & communications	706.64	1,313.56	2,595.30	2,813.31	2,970.86	3,193.67	3,366.13
Transport & storage	627.12	1,119.18	2,150.12	2,317.83	2,477.76		
Financing, insurance, real estate & business services	629.45	943.54	2,189.18	1,946.18	2,049.33	2,143.60	2,456.57
Ownership of dwellings	639.48	752.16	921.20	940.55	996.98	1,046.83	1,099.17
Public administration & defense	1,343.99	1,422.92	1,646.29	1,672.63	1,712.77	1,729.90	1,728.17
Community, social & personal services	1,722.17	2,269.20	2,912.30	3,031.70	3,195.41	3,361.57	3,536.37
Education	859.14	1,131.93	1,437.57	1,532.45	1,612.14		
Gross Domestic Product at F.C.	15,818.17	23,835.71	33,484.64	35,872.01	37,880.07	41,424.30	44,510.13
Growth Rate (% p.a. at Constant Prices)							
Agriculture, forestry & fishing		4.9	-21.7	24.5	4.1	6.3	0.3
Fishing		-2.1	3.2	15.3	1.7		
Mining & quarrying		15.9	-8.5	16.1	-3.8	9.2	8.4
Manufacturing		11.9	-1.1	7.2	4.0	10.9	14.6
Electricity, gas & water		19.2	7.5	9.7	8.4	13.1	10.3
Electricity & gas		18.8	7.5	9.4	8.5		
Construction		13.8	-1.4	-5.4	20.0	21.2	-0.1
Wholesale & retail trade, restaurants & hotels		6.7	-3.6	6.4	4.5	11.2	10.6
Transport, storage & communications		10.7	2.5	8.4	5.6	7.5	5.4
Transport & storage		9.7	0.8	7.8	6.9		
Financing, insurance, real estate & business services		6.2	23.5	-11.1	5.3	4.6	14.6
Ownership of dwellings		2.6	3.6	2.1	6.0	5.0	5.0
Public administration & defense		1.9	3.7	1.6	2.4	1.0	-0.1
Community, social & personal services		3.1	3.8	4.1	5.4	5.2	5.2
Education		4.7	5.8	6.6	5.2		
Gross Domestic Product at F.C.		7.4	-3.3	7.1	5.6	9.4	7.4

/a Old SNA figures.

Source: Data provided by Bank of Korea.

Table A3.7: COMPOSITION AND GROWTH OF DOMESTIC CAPITAL FORMATION BY TYPE OF CAPITAL GOODS

	1980	1981	1982	1983	1984	1985P
Composition (%, current prices)						
Fixed capital formation	100.4	94.5	106.7	104.9	98.1	98.0
Dwellings	18.5	14.7	17.8	19.6	16.1	15.6
Nonresidential buildings	17.2	15.3	18.7	19.6	19.5	18.0
Other construction except land improvement	18.4	18.9	22.8	21.6	20.4	21.6
Land improvement, plantation and orchard development	4.8	5.3	5.5	5.1	4.9	4.3
Transport equipment	11.2	12.3	11.6	10.9	10.7	10.0
Machinery	30.2	27.9	29.8	27.4	25.9	28.3
Livestock	0.1	0.2	0.5	0.6	0.6	0.2
Increase in stock	-0.4	5.5	-6.7	-4.9	1.9	2.0
Gross Capital Formation	100.0	100.0	100.0	100.0	100.0	100.0
Growth Rate (%, 1980 constant prices)						
Fixed Capital Formation		-4.0	12.9	17.1	10.7	2.8
Dwellings		-19.3	22.4	32.5	-4.1	1.1
Nonresidential buildings		-9.3	24.0	25.8	15.8	-4.4
Other construction except land improvement		3.7	23.2	13.8	10.4	10.1
Land improvement, plantation and orchard development		11.5	2.0	13.3	10.1	-7.4
Transport equipment		6.5	-5.5	8.8	16.8	-1.4
Machinery		-2.8	4.3	8.8	15.6	7.0
Livestock		9.7	105.0	16.3	19.3	-1.8
Gross Capital Formation		6.3	0.1	17.5	18.6	1.6

P = Preliminary.

Source: Bank of Korea, New System of National Accounts, 1986.
Bank of Korea, Preliminary Estimates of 1985 National Accounts, 1986.

Table A3.8: COMPOSITION AND GROWTH OF DOMESTIC CAPITAL FORMATION BY TYPE OF PURCHASER /a

	1975	1976	1977	1978	1979	1980	1981	1982	1983	1984
Composition (%, current prices)										
Gross fixed capital formation	84.9	94.0	96.1	98.8	91.9	102.1	99.0	112.1	114.7	103.2
Private enterprises and public enterprises	70.3	78.7	81.2	84.6	78.0	85.2	82.2	93.5	96.3	85.6
Government enterprises	2.8	2.8	2.1	2.1	2.2	2.8	2.7	3.8	3.5	2.1
General government	11.8	12.5	12.8	12.1	11.7	14.1	14.1	14.8	14.8	15.5
Increase in stocks	15.1	6.0	3.9	1.2	8.1	-2.1	1.0	-12.1	-14.7	-3.2
Gross Capital Formation	100.0	100.0	100.0	100.0	100.0	100.0	100.0	100.0	100.0	100.0
(% of GNP)	30.0	25.6	27.7	31.2	35.6	31.3	29.1	27.0	27.8	30.0
Growth Rate (%, 1980 prices)										
Gross fixed capital formation	10.7	19.0	29.0	33.3	8.6	-10.6	-3.3	13.1	16.6	5.7
Private enterprises and public enterprises	6.0	20.9	31.4	36.4	9.3	-12.8	-3.6	13.4	17.3	4.4
Government enterprises	25.1	9.5	-2.6	33.1	16.5	9.3	1.0	43.9	4.5	-29.8
General government	37.6	11.6	21.9	16.3	2.7	1.2	-1.9	5.4	15.8	22.4
Gross Capital Formation	1.7	16.3	23.2	22.8	19.7	-23.7	2.2	5.0	13.7	12.4

/a Old SNA figures.

Source: Bank of Korea, National Income in Korea.

Table A3.9: COMPOSITION AND GROWTH OF GROSS DOMESTIC CAPITAL FORMATION BY INDUSTRIAL USE, 1980-84

	1980	1981	1982	1983	1984
Composition (%, current prices)					
Fixed Capital Formation	100.4	94.5	106.7	104.9	98.1
Agriculture, forestry & fishing	8.1	7.3	7.6	8.0	7.6
Mining & quarrying	0.7	0.8	0.9	0.8	0.7
Manufacturing	22.5	20.0	22.2	19.2	21.4
Electricity, gas & water	9.5	10.2	10.8	10.9	8.2
Construction	2.7	2.7	3.2	2.5	2.5
Wholesale & retail trade, restaurants & hotels	3.6	4.2	6.3	4.0	4.2
Transport, storage & communication	12.3	13.8	13.0	16.3	13.1
Financing, insurance, real estate & business services	24.9	19.5	23.9	25.4	22.9
Social and personal services	2.1	2.0	2.8	2.3	2.5
Government services	14.1	9.3	16.1	15.6	15.0
Increase in stocks	-0.4	0.5	-6.7	-4.9	1.9
Gross Domestic Capital Formation	100.0	100.0	100.0	100.0	100.0
Growth Rate (%, 1980 prices)					
Fixed Capital Formation		-4.0	12.9	17.1	10.7
Agriculture, forestry & fishing		-9.6	0.0	25.7	10.3
Mining & quarrying		16.1	12.8	1.9	12.3
Manufacturing		-7.4	9.8	2.2	33.4
Electricity, gas & water		11.9	4.7	21.1	-10.8
Construction		2.3	17.8	-8.1	22.9
Wholesale & retail trade, restaurants & hotels		18.9	50.0	-23.9	25.7
Transport, storage & communication		11.3	-3.7	48.5	-4.8
Financing, insurance, real estate & business services		-20.0	23.2	27.2	6.2
Social and personal services		-2.4	40.6	-3.7	31.8
Government services		0.5	15.4	16.5	12.3
Increase in stocks		-	-	-	-
Gross Domestic Capital Formation		6.3	0.1	17.5	18.6

Source: Bank of Korea, New System of National Accounts, 1986.
Bank of Korea, Preliminary Estimates of 1985 National Accounts, 1986.

Table A3.10: AVERAGE INCREMENTAL CAPITAL OUTPUT RATIOS, 1970-84 /a

	1971-73	1974-77	1978-81	1982-84
Agriculture, forestry & fishing	1.84	1.73	8.35	3.65
Mining & quarrying	1.49	1.54	8.74	1.56
Manufacturing	1.37	1.40	2.63	1.61
Electricity, gas & water	6.93	10.08	14.82	15.95
Construction	0.58	0.44	2.65	0.98
Wholesale & retail trade, restaurants & hotels	0.71	1.43	4.36	2.62
Transport, storage & communications	7.40	7.46	10.01	15.62
Financing, insurance, real estate & business services	0.87	0.58	0.64	0.70
Ownership of dwellings	27.39	49.75	60.77	46.44
Public administration & defense	11.97	5.38	12.79	46.18
Community, social & personal services	2.00	4.52	6.47	6.75
Gross Domestic Product at F.C.	2.26	2.70	6.54	4.41

/a Changes in output are lagged by one year. Both output and fixed investment are in 1980 constant prices (based on Old SNA figures).

Source: Data provided by Economic Planning Board and Bank of Korea.

Table A4.1: BALANCE OF PAYMENTS SUMMARY, 1976-85
(US$ million)

	1976	1977	1978	1979	1980	1981	1982	1983	1984	1985[P]
Current account	-314	12	-1,085	-4,151	-5,321	-4,646	-2,650	-1,606	-1,371	-882
Trade Balance	-591	-477	-1,781	-4,396	-4,384	-3,628	-2,594	-1,764	-1,036	-30
Exports	7,815	10,047	12,711	14,705	17,214	20,671	20,879	23,204	26,335	26,405
Imports	8,405	10,523	14,491	19,100	21,598	24,299	23,474	24,967	27,371	26,435
Service Balance	-72	266	224	-195	-1,386	-1,518	-554	-435	-876	-1,432
Receipts	1,643	3,027	4,450	4,826	5,363	6,598	7,476	7,179	7,316	6,641
Payments	1,715	2,761	4,226	5,021	6,749	8,117	8,031	7,613	8,192	8,073
Transfers (net)	349	223	472	439	449	501	499	592	541	580
Long-Term Capital (Net)	1,371	1,313	2,166	2,663	1,857	2,842	1,230	1,270	2,000	1,084
Loans and investments	1,639	1,972	2,848	2,833	3,030	3,052	2,896	2,569	2,454	2,238
Amortization	-407	-536	-825	-1,123	-1,085	-1,315	-1,430	-1,672	-1,768	-1,832
Exports on credit	14	-93	-124	-70	-243	57	-350	-677	181	-974
Others	125	-30	267	1,023	155	1,048	114	1,050	1,185	1,652
Basic balance	1,058	1,325	1,081	-1,488	-3,464	-1,804	-1,419	-356	629	201
Short-term capital (net)	357	21	-1,171	844	1,945	-82	4	894	-758	-538
Errors and omissions	-241	-32	-312	-329	-370	-411	-1,296	-942	-889	-878
Overall balance	1,174	1,315	-402	-973	-1,890	-2,297	-2,711	-384	-1,018	-1,215
Bank Borrowing (Net)	364	408	866	1,869	2,861	2,703	3,295	245	1,902	1,742
IMF credit	97	-25	-2	-125	594	628	82	160	319	-235
Bank loans	-39	86	-70	408	130	782	675	480	522	623
Refinance	-131	-3	522	811	742	1,124	1,145	-298	-274	-54
Others	422	350	416	775	1,395	169	1,393	-97	1,335	1,418
Change in foreign exchange reserves	1,419	1,346	631	771	863	320	93	-74	740	99
Foreign exchange reserves	2,961	4,306	4,937	5,708	6,571	6,891	6,984	6,910	7,650	7,749

P = preliminary.

Source: Bank of Korea.

Table A4.2: INDICES OF FOREIGN TRADE AND TERMS OF TRADE, 1970-85
(1980 = 100)

Year	Quantum index		Unit value index		Net barter /a terms of trade	Income terms of trade /b
	Exports	Imports	Exports	Imports		
1970	12.0	29.9	39.9	29.8	133.8	16.1
1971	15.5	36.0	39.4	29.7	132.7	20.6
1972	23.3	37.4	39.9	30.2	132.1	30.8
1973	36.5	47.0	50.4	40.2	125.4	45.8
1974	40.0	49.1	63.8	62.5	102.1	40.8
1975	49.0	50.4	59.2	64.3	92.1	45.1
1976	66.5	62.4	66.2	63.0	105.1	69.9
1977	79.2	75.2	72.4	64.4	112.4	89.0
1978	90.6	98.4	80.1	68.0	117.8	106.7
1979	89.7	110.0	95.8	83.1	115.3	103.4
1980	100.0	100.0	100.0	100.0	100.0	100.0
1981	117.5	111.1	103.2	105.4	97.9	115.0
1982	125.1	111.4	99.7	97.6	102.2	127.9
1983	145.5	126.2	95.9	93.0	103.1	150.0
1984	168.2	145.9	99.2	94.2	105.3	177.1
1985	181.0	154.8	95.5	90.2	105.9	191.7

/a Defined as export price/import price ratio.

/b Export quantum multiplied by net barter terms of trade.

Source: Economic Planning Board, Economic Statistics Yearbook.

Table A4.3: KOREA: INDICES OF EXTERNAL COMPETITIVENESS /a
(1980 = 100)

Period		Nominal effective exchange rate /b (1)	Relative prices /c (2)	Real effective exchange rate (1) x (2)
1981	I	90.9	107.4	98.0
	II	93.5	109.6	102.8
	III	96.7	112.6	109.2
	IV	94.2	112.0	105.8
1982	I	94.1	110.9	104.7
	II	94.3	110.8	104.8
	III	96.0	110.3	106.2
	IV	97.1	109.8	106.9
1983	I	93.6	110.2	103.4
	II	93.6	108.9	102.2
	III	94.0	107.6	101.5
	IV	93.3	106.3	99.5
1984	I	93.5	105.3	98.8
	II	94.0	104.5	98.5
	III	97.0	103.4	100.6
	IV	98.2	102.7	101.2
1985	I	100.0	101.5	101.8
	II	95.4	100.7	96.4
	III	91.1	100.4	91.7
	IV	85.4	100.5	86.1

/a An increase in the index indicates an appreciation of the won.
/b Trade-weighted.
/c Ratio of Korean consumer prices to consumer prices of trading partners.
/d Nominal exchange rate adjusted for relative price changes.

Source: Data provided by Korean authorities.
Staff calculation.

Table A4.4: KOREA: EXPORTS AND EXPORT ORDERS, 1978-85
(In % change from preceding year)

		Exports		Export orders /a	
		Value	Volume	Value	Volume
1978	I	28.3	19.2	19.3	10.8
	II	24.5	14.6	18.8	9.2
	III	26.2	12.8	30.0	16.2
	IV	27.2	16.6	19.3	2.0
	Year	26.5	14.4	21.7	10.0
1979	I	17.4	-3.5	24.3	2.4
	II	19.7	-1.5	12.9	-7.2
	III	24.4	3.1	5.1	-12.9
	IV	13.1	-0.7	-1.3	-13.4
	Year	18.5	-1.0	9.8	-8.2
1980	I	21.6	13.2	14.7	6.3
	II	14.3	9.4	9.7	4.7
	III	12.0	7.4	25.6	20.0
	IV	18.1	18.4	49.3	49.7
	Year	16.3	11.5	24.4	19.2
1981	I	22.5	16.9	24.8	19.5
	II	30.8	23.4	30.0	23.0
	III	23.6	20.0	2.0	-0.9
	IV	10.7	7.3	-0.9	-4.0
	Year	21.4	17.5	13.0	9.5
1982	I	5.9	6.5	-7.5	-7.0
	II	1.9	5.3	-9.0	-10.7
	III	2.8	7.7	-8.8	-4.5
	IV	1.2	7.0	-14.9	-9.9
	Year	2.8	6.5	-10.1	-6.9
1983	I	-1.4	6.9	3.0	11.7
	II	10.9	15.6	3.0	7.4
	III	11.2	14.3	13.7	16.9
	IV	24.9	24.8	19.0	18.9
	Year	11.9	16.3	9.2	13.5
1984	I	31.3	27.3	18.4	14.6
	II	22.0	17.9	16.9	12.9
	III	11.0	6.4	3.4	-0.9
	IV	17.4	14.4	1.9	-0.8
	Year	19.6	15.6	10.3	6.7
1985	I	-8.3	-8.2	-9.7	-9.5
	II	-0.6	2.5	-4.5	-1.4
	III	4.8	11.3	8.5	15.1
	IV	15.4	22.3	12.1	n.a.
	Year	3.6	7.6	0.9	n.a.

/a Arrivals of export letters of credit.

Sources: Bank of Korea, Economic Statistics Yearbook (various issues); data provided by the Korean authorities; and staff estimates.

Table A4.5: MERCHANDISE EXPORTS BY COMMODITY GROUP, 1976-85
(US$ million)

	1976	1977	1978	1979	1980	1981	1982	1983	1984	1985
Food & live animals	508	945	933	1,082	1,153	1,323	1,081	1,092	1,149	1,136
Beverages & tobacco	78	108	120	118	124	119	128	126	119	107
Crude materials, inedible except fuels	196	300	329	361	333	284	275	292	328	298
Mineral fuels, lubricants & related materials	145	117	41	18	33	159	286	536	804	951
Animal & vegetable oils & fats	1	5	11	27	13	15	9	4	4	4
Chemicals	120	226	341	532	780	682	740	747	918	936
Manufactured goods by materials	2,337	3,019	3,784	4,815	6,236	7,215	6,631	6,940	7,353	7,064
Machinery & transport equipment	1,280	1,741	2,587	3,102	3,451	4,712	6,042	7,872	10,322	11,384
Miscellaneous manufactured articles	3,028	3,544	4,536	4,980	5,298	6,638	6,616	6,797	8,213	8,372
Not elsewhere classified	23	41	30	22	85	107	47	38	34	32
Total	7,715	10,047	12,711	15,056	17,505	21,254	21,853	24,445	29,245	30,283

Source: The Bank of Korea, Monthly Statistical Bulletin.

Table A4.6: MERCHANDISE EXPORTS BY PRINCIPAL COUNTRIES, 1976-85

	1976	1977	1978	1979	1980	1981	1982	1983	1984	1985
					US$ million					
United States	2,493	3,119	4,058	4,374	4,607	5,661	6,243	8,245	10,479	10,754
Japan	1,802	2,148	2,627	3,353	3,039	3,503	3,388	3,404	4,602	4,543
Hongkong	325	342	385	531	823	1,155	904	818	1,281	1,566
Indonesia	49	69	103	195	366	370	383	252	254	196
United Kingdom	254	304	393	542	573	705	1,103	1,005	956	913
Germany	398	480	663	845	875	804	758	775	924	979
Others	2,394	3,585	4,482	5,216	7,222	9,056	9,074	9,946	10,749	11,332
Total	7,715	10,047	12,711	15,056	17,505	21,254	21,853	24,445	29,245	30,283
					% of total					
United States	32.3	31.0	31.9	29.1	26.3	26.6	28.6	33.7	35.8	35.5
Japan	23.4	21.4	20.7	22.3	17.4	16.5	15.5	13.9	15.7	15.0
Hongkong	4.2	3.4	3.0	3.5	4.7	5.4	4.1	3.3	4.4	5.2
Indonesia	0.6	0.7	0.8	1.3	2.1	1.7	1.8	1.0	0.9	0.6
United Kingdom	3.3	3.0	3.1	3.6	3.3	3.3	5.0	4.1	3.3	3.0
Germany	5.2	4.8	5.2	5.6	5.0	3.8	3.5	3.2	3.2	3.2
Others	31.0	35.7	35.3	34.7	41.2	42.6	41.5	40.7	36.8	37.4

Source: Economic Planning Board, Major Statistics of Korean Economy.

Table A4.7: KOREA'S EXPORT SHARE BY IMPORT COUNTRY
(%)

	1977	1978	1979	1980	1981	1982	1983	1984	1985
United States	1.5	1.7	1.9	1.8	1.9	2.2	2.2	2.9	3.0
Japan	3.0	3.3	3.1	2.2	2.4	2.5	2.7	3.1	3.2
Hong Kong	3.5	2.8	2.9	3.5	4.0	3.2	2.9	3.3	3.6
United Kingdom	0.5	0.5	0.6	0.5	0.6	0.6	0.5	0.6	0.6
Germany	0.5	0.6	0.5	0.5	0.6	0.5	0.6	0.5	0.5
Industrial countries	1.1	1.3	1.4	1.1	1.3	1.2	1.4	1.6	1.5
Non-oil developing countries	0.3	0.4	0.6	0.6	0.8	1.0	1.2	1.4	1.5
Oil exporting countries	1.5	1.7	1.7	1.9	2.0	2.1	2.0	2.2	2.0
World	1.0	1.4	1.3	1.2	1.4	1.4	1.5	1.7	1.6

Source: IMF, Direction of Trade Statistics.

Table A4.8: MERCHANDISE IMPORTS BY COMMODITY GROUP, 1976-85

	1976	1977	1978	1979	1980	1981	1982	1983	1984	1985
Food & live animals	627	715	931	1,432	1,789	2,721	1,561	1,712	1,622	1,401
Beverages & tobacco	30	34	52	71	85	68	10	30	65	50
Crude materials, inedible except fuels	1,565	1,941	2,395	3,260	3,634	3,630	3,370	3,480	3,951	3,875
Mineral fuels, lubricants & related materials	1,747	2,179	2,453	3,779	6,638	7,765	7,593	6,958	7,274	7,365
Animal & vegetable oils & fats	63	86	104	152	119	137	137	141	173	146
Chemicals	866	1,005	1,298	2,009	1,836	2,109	2,084	2,279	2,762	2,792
Manufactured goods by materials	1,146	1,518	2,225	2,722	2,436	2,775	2,610	3,000	3,762	3,557
Machinery & transport equipment	2,387	2,908	4,947	6,125	4,977	6,000	6,009	7,548	9,797	10,467
Miscellaneous manufactured articles	333	411	557	718	686	787	784	948	1,119	1,248
Not elsewhere classified	10	13	11	71	93	141	93	96	101	234
Total	8,774	10,811	14,972	20,339	22,292	26,131	24,251	26,192	30,631	31,136

Source: The Bank of Korea, Monthly Statistical Bulletin.

Table A4.9: MERCHANDISE IMPORTS BY PRINCIPAL COUNTRIES, 1976-85

	1976	1977	1978	1979	1980	1981	1982	1983	1984	1985
					US$ million					
United States	1,963	2,447	3,043	4,603	4,890	6,050	5,956	6,274	6,876	6,489
Japan	3,099	3,927	5,981	6,657	5,858	6,375	5,305	6,238	7,640	7,560
Hongkong	37	36	51	88	98	201	244	221	468	493
Indonesia	239	354	408	592	485	385	683	387	653	669
United Kingdom	171	148	211	499	304	398	403	468	565	566
Germany	238	347	491	844	637	672	680	650	795	979
Others	3,027	3,552	4,787	7,056	10,020	12,050	10,980	11,954	13,634	14,380
Total	8,774	10,811	14,972	20,339	22,292	26,131	24,251	26,192	30,631	31,136
					% of total					
United States	22.4	22.6	20.3	22.6	21.9	23.2	24.6	24.0	22.4	20.8
Japan	35.3	36.3	40.0	32.7	26.3	24.4	21.9	23.8	24.9	24.3
Hongkong	0.4	0.3	0.3	0.4	0.4	0.8	1.0	0.8	1.5	1.6
Indonesia	2.7	3.3	2.7	2.9	2.2	1.5	2.8	1.5	2.1	2.1
United Kingdom	2.0	1.4	1.4	2.5	1.4	1.5	1.7	1.8	1.8	1.8
Germany	2.7	3.2	3.3	4.2	2.9	2.6	2.8	2.5	2.6	3.1
Others	34.5	32.9	32.0	34.7	44.9	46.1	46.9	45.6	44.5	46.3

Source: Economic Planning Board, *Major Statistics of Korean Economy*.

Table A4.10: IMPORTS OF LIBERALIZED ITEMS

	Imports of liberalized items			
	Total	For export	For domestic	Total imports
Items Liberalized in July 1983 (305 items)				
83.7-83.12				
Value (US$ million)	500	276	224	14,029
Growth rate (%)	11.9	7.3	23.1	12.9
83.7-84.6				
Value (US$ million)	1,215	939	276	29,515
Growth rate (%)	34.8	45.5	7.8	20.1
84.7-85.6				
Value (US$ million)	1,335	1,110	225	29.470
Growth rate (%)	9.9	18.2	-18.3	-0.2
85.1-85.6				
Value (US$ million)	643	522	120	14.326
Growth rate (%)	-10.0	-12.5	2.2	-7.5
84.7-85.11				
Value (US$ million)	1,790	1,467	323	42,517
Growth rate (%)	-0.4	2.2	-10.7	1.0
Items Liberalized in July 1984 (326 items)				
84.7-84.12				
Value (US$ million)	439	226	213	15,145
Growth rate (%)	31.3	41.5	21.9	7.9
84.7-85.6				
Value (US$ million)	876	421	456	29,470
Growth rate (%)	1.9	-0.6	9.9	-0.2
85.1-85.6				
Value (US$ million)	437	195	243	14,326
Growth rate (%)	-16.7	-31.8	1.2	-7.5
84.7-85.11				
Value (US$ million)	1,245	610	635	42,57
Growth rate (%)	1.7	-3.3	8.2	1.0

Source: Data provided by Economic Planning Board

Table 4.11: TARIFF REVENUE COMPARISONS
(W millions)

	1981		1982		1983		1984	
	Tariff revenue	Effective tariff rate /a	Tariff revenue	Effective tariff rate	Tariff revenue	Effective tariff rate	Tariff revenue	Effective tariff rate
Live animals	20,556	22.32	35,366	19.83	37,243	18.25	18,485	17.75
Vegetable products	82,860	5.41	68,306	8.06	81,618	8.05	108,631	9.78
Animal vegetable fats and oils	8,670	10.01	11,485	12.32	16,774	15.49	21,277	15.24
Prepared foodstuffs	60,286	23.72	43,892	24.21	69,221	27.91	56,277	23.15
Mineral products	28,186	0.52	37,236	0.69	180,621	3.55	245,005	4.59
Chemical products	123,584	17.91	142,962	20.97	176,800	22.58	182,726	18.30
Artifical resins & plastic materials	40,179	24.06	48,366	24.77	65,779	24.40	63,081	22.33
Raw hides and skins, leathers products	3,218	19.51	4,777	19.14	6,136	22.27	5,380	16.51
Wood and wood products	10,865	3.46	11,924	3.20	18,391	4.05	19,835	4.11
Paper and paper products	23,464	8.42	21,457	9.84	30,126	11.58	12,967	9.18
Textiles	18,414	11.65	24,449	18.10	29,191	15.48	28,421	16.77
Footwear and other manufactured products	324	47.16	408	27.79	682	43.84	1,024	31.40
Stone and cements products	8,720	28.11	24,402	26.18	33,005	28.82	34,385	28.41
Pearls and precious and semiprecious products	683	6.73	1,994	9.81	3,170	10.71	2,401	11.90
Metal products	31,814	10.67	83,652	12.39	104,506	12.55	123,763	13.79
Machinery and elec. equipment	183,669	8.20	212,822	8.34	290,619	9.52	380,993	12.95
Vehicles, aircraft and transport equipment	41,936	3.97	41,827	4.10	50,998	2.99	48,255	7.65
Optical and audiovisual products	34,035	14.96	42,007	13.04	62,738	14.15	78,148	14.86
Arms and ammunition	28	0.09	54	0.25	187	0.79	70	0.19
Miscellaneous manufactured articles	2,494	25.93	5,284	30.00	8,853	30.72	8,698	25.66
Works of arts and collectors' pieces	-	-	-	-	-	-	-	-
Total	890,584	5.92	963,453	7.33	1,278,531	8.57	1,482,845	10.19

/a Defined as collections (excluding rebates) as a proportion of total imports for domastic use.

Source: Data provided by MOF.

Table A5.1: OUTSTANDING EXTERNAL DEBT BY MATURITY AND BORROWER, 1977-85

	1978	1979	1980	1981	1982	1983	1984	1985
Debt with maturity of more than three years	10,926	14,132	16,327	19,899	22,426	24,999	28,042	32,176
Financial institutions	620	1,980	2,039	4,174	5,332	6,867	8,314	10,179
Public sector	4,340	5,271	6,505	7,862	9,342	10,292	11,056	11,376
Private sector	5,045	5,603	6,177	6,440	6,310	6,155	5,905	5,742
Others	921	1,298	1,606	1,423	1,442	1,685	2,767	4,901
Debt with maturities of one to three years	483	561	754	1,061	971	1,910	2,016	2,346
Short-term debt of the private sector	1,041	2,251	4,264	4,193	4,159	4,997	4,126	3,640
Trade credits	840	1,811	3,436	3,454	3,339	4,114	3,025	2,727
Borrowing for oil bill	-	-	367	371	447	584	804	637
Others	201	440	561	368	373	299	297	276
Subtotal	12,450	16,944	21,345	25,153	27,556	31,906	34,184	38,162
Short-term debt of financial sector	2,110	3,205	5,112	6,034	8,268	7,118	7,299	7,092
Refinanced	1,216	2,028	2,770	3,892	5,038	4,740	4,444	3,903
Deposits	128	143	200	222	145	163	218	298
Others	218	229	447	287	1,092	741	839	1,154
Interoffice "A" account of foreign bank branches	558	805	1,685	1,633	1,993	1,474	1,798	1,737
Use of fund credit	263	138	713	1,246	1,259	1,354	1,570	1,538
Total	14,823	20,287	27,170	32,433	37,083	40,378	43,053	46,762
Memorandum item								
Undisbursed loans with maturity of more than three years /a	3,020	4,191	3,332	5,025	4,252	4,298	n.a.	n.a.

/a Public and commercial loan and Bank loan.

Source: Ministry of Finance.

Table A5.2: EXTERNAL DEBT INDICATORS, 1977-85

	1977	1978	1979	1980	1981	1982	1983	1984	1985
				US$ billion					
Nominal stock of debt, end of period	12.6	14.8	20.4	27.3	32.5	37.1	40.4	43.1	46.8
(Short term external debt)		(3.2)	(5.5)	(9.4)	(10.2)	(12.4)	(12.1)	(11.4)	(10.7)
Stock of debt deflated by:									
Export unit prices (1977=100)	12.6	13.4	15.5	19.7	22.8	27.3	30.5	31.5	35.5
Import unit prices (1977=100)	12.6	14.1	15.9	16.7	19.0	23.6	28.0	29.5	30.8
				%					
Ratio of debt /a to GNP	31.0	26.5	28.4	39.1	49.1	53.5	53.2	52.3	56.3
Ratio of gross international reserves to debt	34.0	33.2	27.8	24.0	21.2	18.7	18.5	19.1	17.9
Ratio of total debt /a to exports of goods and services	88.7	80.2	90.6	106.0	109.8	123.1	132.9	128.1	142.2
Ratio of stock of short-term debt /b to imports of goods and services	13.8	13.9	19.3	26.7	26.1	32.3	37.1	32.1	32.0

/a Average stock during the year.

/b Excludes interoffice "A" account of foreign bank branches.

Sources: Ministry of Finance; staff calculations.

Table A5.3: DEBT SERVICE,/a 1977-85
(US$ million)

	1977	1978	1979	1980	1981	1982	1983	1984	1985
Interest /b									
Medium- & long-term debt	566	758	994	1,426	1,896	2,386	2,226	2,532	2,692
Short-term debt	116	214	471	1,165	1,580	1,234	961	1,288	995
Subtotal	682	972	1,465	2,591	3,476	3,620	3,187	3,820	3,687
Amortization /c									
Financial institutions	143	339	191	159	210	282	401	582	810
Public sector	100	138	173	257	328	384	493	660	704
Private sector	431	685	1,023	834	987	1,046	1,129	1,108	1,128
Medium-term trade liability	150	201	272	329	474	489	510	681	826
Subtotal	824	1,363	1,659	1,579	1,999	2,201	2,533	3,031	3,468
Total Debt Service	1,506	2,335	3,124	4,170	5,475	5,821	5,720	6,851	7,155
Debt service ratio of MLT /d	10.6	12.3	13.6	13.3	14.3	16.2	15.7	16.6	18.7
Debt service ratio (%) /e	11.5	13.6	16.0	18.5	20.1	20.6	18.8	20.4	21.7
Interest payments ratio (%) /f	5.2	5.7	7.5	11.5	12.8	12.8	10.5	11.4	11.2
Memorandum Item									
Exports of goods and services	13,074	17,161	19,531	22,577	27,250	28,323	30,348	33,565	32,931

/a Includes IMF.
/b Excludes interest on medium-term liability which is included in amortization.
/c Excludes amortization of debt with maturities of less than one year.
/d Ratio of medium- and long-term debt service to exports of goods and services.
/e Ratio of total debt service to exports of goods and services.
/f Ratio of interest payments to exports of goods and services.

Source: Ministry of Finance; staff calculations.

Table A5.4: DEBT SERVICE PROJECTIONS, 1985-91
(US$ million)

	1985	1986	1987	1988	1989	1990	1991
Debt Service							
Principal (1)	3,687	4,000	4,600	5,050	5,550	5,900	6,200
Interest (2)	3,740	3,920	4,220	4,370	4,455	4,540	4,590
On medium- and long-term debt (a)	2,692	2,780	2,960	3,090	3,220	3,230	3,130
On short-term debt (b)	995	940	980	1,000	1,030	1,010	990
Current Receipts (c)							
Exports	26,441	30,800	34,400	38,400	43,000	47,800	53,100
Invisible	7,300	7,800	8,700	10,600	10,600	11,500	12,500
(1 + 2) / c (%)	22.0	20.5	20.5	19.2	18.7	17.6	16.4
(1 + 2a) / c (%)	18.9	17.6	17.5	16.6	16.4	15.4	16.5

Source: Data provided by Economic Planning Board

Table A5.5: GROSS ANNUAL EXTERNAL FINANCING
(US$ million)

	1981	1982	1983	1984	1985
Long-term	6,898	5,127	6,975	7,954	8.353
Loans and investment	3,052	2,897	2,568	2,458	2,238
(Public loans)	(1,705)	(1,877)	(1,494)	(1,424)	(1,024)
(Commercial Loans)	(1,242)	(919)	(973)	(858)	(964)
(Investment)	(105)	(101)	(101)	(171)	(250)
Borrowings by financial Institutions	2,400	1,731	2,731	3,858	4,351
(B.L.)	(2,079)	(1,457)	(1.826)	(2,054)	(2,368)
(Bond)	(90)	(44)	(133)	(585)	(1,219)
(FRCD)	30)	(70)	(140)	(245)	(370)
(P/N)	(201)	(160)	(632)	(974)	(574)
IMF facilities	677	121	204	587	132
Others	769	378	1,472	1,056	1,076
Short-term	840	2,478	-256	-690	-693
Private sector	-82	4	894	-871	-486
Monetary sector	922	2,474	-1,150	181	-207

Source: Data provided by Ministry of Finance.

Table A5.6: FOREIGN INVESTMENT BY AREA OF ACTIVITY
(US$'000)

	1962-66		1967-71		1972-76		1977-81		1982		1983		1984		1985		Total		
	No.	Amount	No.	Amount	No.	Amount	No.	Amount	No.	Amount	No.	Amount	No.	Amount	No.	Amount	No.	Amount	%
Agriculture and Fisheries	1	102	8	943	19	5,221	11	6,350	2	1,250	1	1,149	3	600	4	3,650	49	19,265	0.7
Agriculture	1	102	6	869	6	1,478	8	3,323	1	150	1	770	1	225	2	3,335	26	10,253	0.4
Fisheries	-	-	2	74	13	3,743	3	3,026	1	1,100	-	379	2	375	2	315	23	9,012	0.3
Mining and Manufacturing	14	22,897	139	59,925	398	427,338	153	223,308	45	122,342	58	102,879	86	264,285	100	181,858	993	1,574,854	59.3
Mining	-	-	-	-	7	1,451	4	812	2	554	1	270	1	520	3	816	17	4,423	0.2
Manufacturing	14	22,897	139	59,925	391	425,937	149	392,468	44	121,788	57	102,609	85	263,765	97	181,042	976	1,570,431	59.2
Foodstuffs	1	100	4	1,056	3	1,729	10	29,276	4	10,646	5	5,928	10	18,426	9	3,628	46	70,789	2.7
Textile and garments	3	759	15	4,048	44	57,505	5	2,562	-	2,987	-	1,940	5	1,941	1	807	73	72,549	2.7
Paper and wood products	-	-	-	-	2	263	1	850	-	-	-	-	-	-	-	-	3	1,113	0.0
Chemical products	-	-	20	11,626	56	146,143	24	120,258	8	41,321	5	7,594	6	6,703	17	44,773	136	378,418	14.3
Medical products	3	313	4	2,654	3	2,082	4	9,377	10	24,162	8	20,784	4	6,399	6	11,268	42	77,039	2.9
Fertilizer	2	21,500	-	-	2	21,325	-	150	-	-	-	-	-	-	-	-	4	41,975	1.6
Petroleum	-	-	3	7,845	1	15,407	-	8,283	-	-	-	-	1	5,000	-	-	5	36,535	1.4
Ceramics	-	-	7	5,649	12	4,389	3	4,231	-	128	3	654	1	4,526	3	8,924	29	28,491	1.0
Metal products	-	-	13	8,975	33	30,003	17	21,318	4	2,936	5	1,788	5	5,058	6	2,570	83	72,648	2.7
Machinery	1	20	19	5,287	62	30,330	37	39,639	2	9,824	9	5,725	20	131,498	20	50,710	170	277,533	10.5
Electrical and electronics	3	901	27	10,614	121	82,995	21	107,765	6	19,212	13	41,368	21	67,508	26	55,806	238	386,169	14.5
Transport equipment	-	-	1	40	4	25,542	2	19,485	1	7,664	1	13,835	-	-	-	-	9	66,566	2.5
Miscellaneous products	1	304	26	2,131	48	8,234	25	23,774	9	2,908	8	2,993	12	16,706	9	2,556	138	60,606	2.3
Social Overhead Capital	-	-	17	11,805	33	132,630	33	188,113	8	64,199	16	163,725	14	154,164	23	346,212	144	1,060,848	40.0
Financing	-	-	2	1,621	3	11,340	5	52,472	1	21,971	-	3,326	2	13,266	4	17,267	17	121,263	4.6
Construction	-	-	7	3,307	9	15,856	11	37,304	4	12,162	10	3,895	8	22,457	13	16,147	62	111,168	4.2
Electricity	-	-	-	-	-	-	2	8,395	-	-	-	-	-	-	-	-	2	3,395	0.1
Transportation and storage	-	-	5	783	4	3,412	4	24,437	-	2,000	2	784	-	62	-	509	15	31,987	1.2
Hotel and tourism	-	-	3	6,094	17	102,022	11	70,465	3	28,066	4	155,720	4	118,379	6	312,289	48	793,035	29.9
Total	15	22,999	164	72,673	450	565,239	197	587,743	55	187,791	75	267,753	103	419,049	127	531,720	1,186	2,654,967	100.0

Source: Data provided by Economic Planning Board.

Table A6.1: PUBLIC SECTOR RESOURCE BALANCE
(% of GNP)

	Central Government /a				
	1977-81 Actual	1982 Actual	1983 Actual	1984 Actual	1985 Budget
Current revenue /c	21.0	22.5	21.3	21.3	21.0
Tax revenue	16.3	17.5	17.5	16.9	17.3
Transfer from other levels of Government	-	-	-	-	-
Current expenditures /c	16.3	20.1	17.6	17.8	17.8
Current balance	4.1	2.4	3.7	3.5	3.1
Gross fixed investment	1.8	4.0	1.8	1.6	1.8
Overall balance (- deficit)	-3.0	-4.6	1.6	1.4	1.4

/a Includes nonfinancial public enterprises other than communications.

/b General account only.

/c Plan figures for current revenues and expenditures are not strictly comparable owing to differences in the consolidation of subaccounts in the budget.

Source: Economic Planning Board.

Table A6.2: KOREA: SUMMARY OF PUBLIC SECTOR REVENUES AND EXPENDITURES
(Won billion)

	1983 Actual	1984 Budget	1984 Actual	1985 Budget	1985 Actual
Current revenue	12,436	12,236	13,958	14,593	15,261
Current expenditure	10,295	10,028	11,675	11,585	11,957
Savings	2,141	2,183	2,283	3,008	3,304
	(3.6)	(3.3)	(3.4)	(4.2)	(4.6)
Capital revenue	168	183	136	254	279
Capital expenditure and net lending	3,260	2,753	3,419	4,386	4,383
Capital balance	3,092	2,570	3,283	4,132	4,104
Overall balance	951	387	-1,001	-1,124	-800
	(1.6)	(0.6)	(-1.5)	(-1.6)	(-1.1)
Financing	951	387	1,001	1,124	800
Domestic	552	685	691	664	360
Foreign	399	298	310	460	440
	(0.7)	(0.4)	(0.5)	(0.6)	(0.6)
Memorandum items					
Gross fixed capital formation	1,039	1,013	1,075	1,212	
	(1.8)	(1.5)	(1.6)	(1.7)	n.a.

Notes: (1) Based on data on consolidated central government and nonfinancial public enterprises excluding local government.

(2) Figures in parenthese are ratios to GNP (current market prices).

Source: Ministry of Finance.

Table A6.3: NET EXPENDITURE BY VARIOUS GOVERNMENT SPECIAL FUNDS/ACCOUNTS
(Billion won)

	1976	1977	1978	1979	1980	1981	1982	1983	1984	1985 Budget
Enterprise Special Accounts										
Grain management	3.8	4.6	6.3	7.2	9.1	10.6	11.3	11.6	11.4	15.6
Monopoly	190.1	226.4	296.4	383.1	456.1	720.8	803.7	864.9	854.2	850.4
Railway services	81.6	83.1	102.4	161.2	203.4	179.8	209.0	186.9	218.5	271.7
Communications	122.1	156.6	223.9	253.8	676.5	809.3	18.1	29.6	35.0	51.0
Supply	1.3	2.2	1.9	6.9	12.8	9.5	7.5	10.0	11.5	8.9
Special Accounts										
Loan management						330.5	343.6	390.9	324.1	495.9
Military personnel pension	} 690.1	} 273.9	} 463.7	} 426.8	} 626.2	91.7	119.6	138.5	143.1	185.1
Others	}	}	}	}	}	378.4	403.0	300.4	343.8	423.0
Nonfinancial Public Enterprises										
Grain management fund	203.7	142.4	249.3	228.4	226.0	568.5	523.2	230.6	484.6	100.1
Supply fund	0.6	10.9	25.3	25.5	14.1	2.7	1.1	1.1	0.8	2.2
Funds										
National investment	-	-	402.3	514.7	542.4	723.4	781.1	600.4	411.6	261.8
National housing fund	-	-	-	-	-	594.9	244.7	525.7	968.2	719.7
Others	209.3	260.2	31.2	86.1	171.0	206.8	549.5	182.9	213.5	255.3
Total	1,503.5	1,160.3	1,802.7	2,093.7	2,937.6	4,626.9	4,015.4	3,473.5	4,020.6	3,640.7

Source: Data provided by Economic Planning Board

Table A7.1: KOREA: MONETARY SURVEY, 1978-85

End of period:	1978	1979	1980	1981	1982	1983	1984	1985
(In billions of won)								
Net foreign assets	725	236	-582	-2,264	-4,340	-5,009	-6,095	-7,606
Assets	2,519	2,900	4,806	5,409	5,861	6,461	7,471	8,213
Liabilities	1,794	2,664	5,388	7,673	10,201	11,469	13,566	15,819
Net domestic assets	7,204	9,641	13,117	18,242	24,360	28,810	32,706	39,552
Domestic credit	8,722	11,826	16,778	22,016	27,529	31,847	36,059	42,561
Public sector	(464)	(335)	(731)	(1,742)	(2,158)	(2,013)	(1,973)	(2,013)
Private sector	(8,258)	(11,491)	(16,047)	(20,274)	(25,371)	(29,834)	(34,086)	(40,548)
Net other items	-1,518	-2,185	-3,661	-3,774	-3,169	-3,037	-3,353	-3,009
Broad money	7,929	9,878	12,535	15,671	19,904	22,938	24,706	28,565
Narrow money	2,714	3,275	3,807	3,982	5,799	6,783	6,820	7,558
Quasi-money	5,215	6,603	8,728	11,689	14,105	16,155	17,886	21,007
(Percentage change) /a								
Net foreign assets	-4.3	-6.1	-8.3	-13.5	-13.3	-3.4	-4.7	-24.8
Net domestic assets	39.3	30.7	35.2	39.1	33.5	18.3	13.5	20.9
Domestic credit	46.7	39.1	50.1	41.8	35.2	21.7	18.4	18.4
Public sector	(1.7)	(-1.7)	(4.0)	(8.1)	(2.7)	(-0.7)	(-0.2)	(2.0)
Private sector	(45.0)	(40.8)	(46.1)	(33.7)	(32.5)	(22.4)	(18.5)	(19.0)
Net other items	-7.4	-8.4	-14.9	-3.1	16.0	4.2	-10.4	10.3
Broad money	35.0	24.6	26.9	25.0	27.0	15.2	7.7	15.6

/a Change over preceding 12 months as a percentage of the stock of broad money at the beginning of the period.

Source: Bank of Korea, Monthly Statistics Bulletin.

Table A7.2: SELECTED INTEREST RATES, DEPOSIT MONEY BANKS, 1979-85
(% p.a.)

Effective from:	1979 Sep	1980 Nov	1981 Nov	1982 Nov	1983 Apr	1983 Nov		1984 Jan	1984 Nov	1985 Jul	1985 Oct
Household checking deposits	-/a	-	14.4/b	8.0	8.0	8.0		6.0	6.0	6.0	
Notice deposits (more than 30 days)	10.0	10.5	8.3	-/c	-	-		-	-	-	
Time deposits											
Over 3 months	15.0	14.8	14.8	7.6	7.6	7.6		6.0	6.0	6.0	
Over 6 months	17.1	16.9	15.2	7.6	7.6	7.6		6.0	6.0	6.0	
Over 1 year	18.6	19.5	17.4	8.0	8.0	8.0		9.0	10.0	10.0	
Savings deposits											
Less than 30 days	5.5	5.5	14.4	} 8.0	8.0	8.0		6.0	6.0	6.0	
More than 30 days	12.6	12.3	14.4								
Liberal savings deposits											
Less than 3 months										6.0	
Over 3 months										9.0	
Over 6 months										12.0	
Installment savings deposits, 3 years	18.2	19.5 (22.5)/d	17.3 (20.3)/d	} 8.0	8.0	8.0		9.0	10.0	10.0 (13.0)/d	
Workmen's wealth accumulation deposits /e											
2 years	25.7	28.4	28.4	19.5	19.5	19.5		21.4	21.4	21.4	
5 years	30.2	33.6	33.6	21.0	21.0	21.0		23.9	23.9	23.9	
Certificates of deposits (3-6 months)									11.0	12.0	11.75
Memorandum Items											
Selected interest rates on deposits of nonbank financial institutions											
Investment and financial companies							{ 1-7 days	3.5	3.5	3.5	
Bills resold: 1-29 days	20.6	19.2	18.1	9.0	8.5	8.5	{ 8-29 days	7.0	7.0	7.0	
60-90 days	24.6	23.1	20.0	11.0	10.5	10.5		8.0	8.0	8.0	
Mutual credit cooperatives - time deposits, 1 year	19.8	22.5	19.5	9.0	9.0	9.0		9.0	10.5	11.0	

/a Interest rate of 6.0% between April 1977 and April 1979.
/b Established in June 1982.
/c Abolished from June 1982.
/d Special interest rate on household deposits.
/e These rates include only the interest paid by banks and the government subsidy. In addition, employees are encouraged to contribute to the yield on these assets.

Sources: Bank of Korea, Monthly Statistical Bulletin; and data provided by the Korean authorities.

Table A7.3: SOURCES AND USES OF TOTAL BANK FINANCE

	1978-80		1981-83		1984/85 /a	
	Amount (W bln)	Share (%)	Amount (W bln)	Share (%)	Amount (W bln)	Share (%)
Sources						
Deposits	6,199	46.0	8,932	48.6	2,755	28.6
Foreign liabilities	3,605	26.8	5,240	28.5	2,318	24.1
Credit from Bank of Korea	2,069	15.4	2,327	12.7	3,361	34.9
Public sector deposits	511	3.8	1,406	7.7	478	5.0
Capital accounts	854	6.4	979	5.3	789	8.2
Net other items	218	1.6	-513	-2.8	-67	-0.7
Total	13,461	100.0	18,371	100.0	9,634	100.0
Uses						
Credit to private sector	10,405	77.3	13,756	74.9	6,967	72.3
Foreign assets	1,819	13.5	1,545	8.4	460	4.8
Deposits at Bank of Korea	578	4.3	1,705	9.3	1,689	17.5
Credit to public sector	659	4.9	1,365	7.4	518	5.4
Net other items	-	-	-	-	-	-
Total	13,461	100.0	18,371	100.0	9,634	100.0

/a May 1985.

Source: Economic Planning Board.

Table A7.4: KOREA: MOVEMENTS OF WHOLESALE AND CONSUMER PRICES, 1978-85 (1980=100)
(% changes) /a

	1978	1979	1980	1981	1982	1983	1983 1st qtr	1983 2nd qtr	1983 3rd qtr	1983 4th qtr	1984	1984 1st qtr	1984 2nd qtr	1984 3rd qtr	1984 4th qtr	1985	1985 1st qtr	1985 2nd qtr	1985 3rd qtr	1985 4th qtr
Wholesale Prices																				
All Commodities	11.6	18.8	38.9	20.4	4.7	0.2	0.4	-0.8	-0.3	0.0	0.7	0.4	-0.2	2.3	-0.9	0.9	-0.5	0.5	0.5	0.6
Agricultural & marine foods	33.8	11.4	24.5	27.6	0.3	3.4	4.8	-1.4	-3.1	-2.5	-0.1	2.0	-2.9	14.7	-3.8	6.0	-1.0	1.2	2.0	1.7
Coal & electric power	19.3	33.2	48.4	28.9	10.9	-0.2	0.1	-0.5	0.4	0.0	0.3	0.0	0.6	0.0	0.0	2.1	0.0	2.2	1.0	0.7
Petroleum & related products	3.8	38.4	105.3	34.9	7.8	-5.1	-1.5	-4.0	-0.6	0.2	-1.9	-0.6	-0.3	0.0	-0.4	-0.4	0.0	0.0	0.5	0.1
Producer goods	6.4	24.4	52.0	21.3	5.1	-0.9	-0.2	-1.0	0.1	0.3	0.6	0.2	0.4	0.0	-0.6	-0.2	0.0	0.2	0.0	-0.1
Capital goods	2.2	12.6	19.3	11.5	7.4	1.2	-0.1	0.1	1.3	0.6	1.4	-0.3	0.2	0.2	0.0	0.3	-0.2	0.1	0.3	0.3
Consumer goods	20.4	13.4	27.1	20.4	3.7	1.8	1.4	-0.5	-1.3	-0.5	0.9	0.2	-1.2	6.1	-1.4	2.0	-1.1	0.8	1.1	1.4
Consumer Prices																				
All Items	14.5	18.3	28.7	21.3	7.3	3.4	1.9	0.3	0.1	0.0	2.3	1.9	-0.4	1.7	-0.7	2.5	1.2	0.6	1.1	0.7
Food & beverages	16.7	13.8	26.6	27.5	2.5	1.3	2.7	-0.4	-1.1	-1.0	1.5	3.3	-1.8	3.5	-1.2	3.7	1.9	0.4	2.3	0.4
Excluding food & beverages	12.4	22.6	30.5	18.4	10.9	4.9	1.4	0.8	1.0	0.7	2.9	0.9	0.6	0.4	-0.4	1.6	0.6	0.7	0.4	1.0

/a Annual data are calculated on year-to-year basis; quarterly data are percent change from preceding quarter.

Source: Bank of Korea, Monthly Statistical Bulletin.

Table A9.1: INDUSTRIAL PRODUCTION, 1970-85
(1980 = 100)

	All items		Mining		Manufacturing		Electricity	
	Index	Growth rate (%)	Index	Growth rate (%)	Index	Growth rate (%)	Index	Growth rate (%)
1970	17.8	11.9	64.1	11.7	16.3	11.6	24.6	19.4
1971	20.5	15.2	66.2	3.3	19.0	16.6	28.3	15.0
1972	23.5	14.6	63.1	-4.7	22.1	16.3	31.8	12.4
1973	31.4	33.6	73.4	16.3	30.0	35.7	39.8	25.2
1974	40.0	27.4	80.2	9.3	38.7	29.0	45.2	13.6
1975	47.6	19.0	89.9	12.1	46.3	19.6	53.2	17.7
1976	61.9	30.0	90.6	0.8	61.0	31.7	62.0	16.5
1977	74.1	19.7	98.8	9.1	73.4	20.3	71.3	15.0
1978	91.1	22.9	101.1	2.3	90.9	23.8	84.6	18.7
1979	101.8	11.7	101.1	0.0	101.9	12.1	95.6	13.0
1980	100.0	-1.8	100.0	-1.1	100.0	-1.9	100.0	4.6
1981	112.7	12.7	102.7	2.7	113.4	13.4	108.0	8.0
1982	118.3	5.0	96.9	-5.6	119.4	5.3	115.8	7.2
1983	137.0	15.8	97.5	0.6	139.0	16.4	131.2	13.3
1984	157.5	15.0	104.6	7.3	160.6	15.0	144.5	10.1
1985p	164.3	4.3	113.7	8.7	167.0	4.0	155.8	7.8

p = preliminary

Source: Economic Planning Board.

Table A9.2: CAPACITY UTILIZATION BY INDUSTRIAL CLASSIFICATION
(%)

	1970	1971	1972	1973	1974 /a	1976	1977	1978	1979	1980	1981	1982	1983	1984	1985
Manufacturing	65.9	62.1	65.1	74.3	69.5	78.9	81.7	88.3	82.1	69.5	70.3	69.4	75.8	80.4	79.2
Food and beverages	59.2	61.8	70.6	66.4	62.9	62.4	75.8	85.6	82.4	69.3	64.3	64.9	75.4	78.8	73.0
Textiles and leather	70.4	66.3	76.3	83.0	77.9	87.5	85.9	84.8	82.1	80.1	80.9	80.2	79.0	78.9	76.7
Wood products	n.a.	n.a.	n.a.	n.a.	n.a.	84.6	94.4	99.0	84.6	61.0	59.9	46.6	45.4	51.4	55.4
Paper products	n.a.	n.a.	n.a.	n.a.	n.a.	72.8	80.8	88.4	85.1	75.4	74.8	72.5	76.4	82.1	79.5
Chemical products	61.8	62.1	66.2	73.6	70.0	91.9	98.1	110.4	95.4	80.3	76.0	70.9	75.4	78.5	80.2
Nonmetallic mineral products	72.5	73.1	71.7	75.5	77.2	81.9	88.2	87.3	77.9	63.6	61.1	68.3	77.6	78.4	72.6
Basic metal	60.1	58.9	53.5	63.2	59.0	78.6	81.1	88.1	81.0	71.3	71.2	74.7	83.8	87.3	88.9
Fabricated metal products, machinery and equipment	62.2	59.2	69.4	67.3	67.1	61.0	57.1	61.7	62.6	53.1	61.0	60.0	67.9	77.9	75.5

/a January-June.

Note: Seasonally adjusted figures.

Source: Data provided by Economic Planning Board.

Table A9.3: DEBT-EQUITY RATIOS BY INDUSTRIAL CLASSIFICATION

	1976	1977	1978	1979	1980	1981	1982	1983	1984
Mining	167.3	186.5	257.4	369.6	262.0	295.0	225.9	181.9	137.2
Electricity	133.5	138.8	167.4	197.2	257.9	262.5	158.3	176.9	176.5
Construction	368.7	310.9	349.8	383.5	524.9	522.8	497.2	481.1	450.8
Distribution and hotels	396.1	373.0	431.0	442.1	566.3	590.1	523.2	403.1	465.7
Transport and storage	419.0	474.9	488.5	449.0	628.4	544.7	485.3	496.9	537.1
Real estate and business services	461.8	164.4	185.0	189.5	209.2	183.3	182.1	162.1	285.2
(Other services)	97.8	118.1	141.4	311.0	300.4	252.9	275.5	235.7	196.3
(Manufacturing)	364.6	350.7	366.8	377.1	487.9	451.5	385.8	360.3	392.7
Food and beverages	453.5	369.3	359.5	409.5	471.9	528.8	422.4	395.6	374.5
Textiles, wearing apparel and leather	497.3	500.3	548.6	638.8	820.1	637.7	598.6	579.6	524.8
Wood and furniture	631.0	573.4	560.4	604.7	2,051.4	4,894.8	10,253.0	*	*
Paper printing and publishing	283.2	324.0	307.5	287.5	382.8	483.8	539.4	393.5	337.2
Petrochemical products	242.5	223.2	228.2	299.8	414.7	439.2	367.7	297.0	273.3
Nonmetallic mineral products	323.6	352.2	406.9	422.8	447.9	399.0	290.7	274.4	250.0
Basic metal	342.2	298.0	301.9	324.9	496.5	463.6	243.8	236.9	219.5
Fabricated metal products, machinery and equipment	381.0	447.8	477.7	335.5	403.0	349.6	374.7	375.7	385.7

Legend: * = denominator is negative.

Source: Bank of Korea, _Financial Statement Analysis_.

Table A9.4: PROFITABILITY RATIOS BY INDUSTRIAL CLASSIFICATION

	1976	1977	1978	1979	1980	1981	1982	1983	1984
Mining	10.7	11.1	3.8	0.2	3.4	5.4	5.5	3.8	3.9
Electricity	5.5	7.6	6.6	11.3	10.9	9.0	6.5	6.3	8.4
Construction	15.5	18.4	17.1	13.4	12.2	11.7	10.5	9.7	7.8
Distribution and hotels	13.1	11.3	11.9	12.0	11.5	11.2	8.6	9.2	8.3
Transportation and storage	6.6	8.9	8.6	9.4	8.3	9.9	8.3	4.7	5.5
Real estate and business services	6.7	9.0	8.3	8.6	7.7	11.9	8.7	9.4	5.9
Other services	21.6	15.9	19.3	16.3	16.7	7.6	8.2	11.0	10.4
Manufacturing	10.4	10.8	11.0	10.7	9.1	10.0	8.8	9.6	9.7
Food and beverages	13.2	15.4	16.8	13.4	12.0	11.5	10.6	10.8	8.9
Textiles, wearing apparel and leather	9.6	8.3	11.0	10.4	9.8	11.5	8.3	8.6	8.2
Wood and furniture	7.7	10.6	14.7	8.2	-0.2	5.0	3.3	8.9	4.7
Paper, printing and publishing	13.3	14.5	14.2	14.5	10.8	7.4	8.7	10.0	10.0
Petrochemical products	13.2	10.5	14.1	14.0	12.2	10.9	10.8	11.0	12.4
Nonmetallic mineral products	10.2	12.8	10.5	12.0	10.5	9.4	9.6	11.7	12.2
Basic metal	7.1	9.2	7.7	8.4	6.2	9.6	6.6	8.4	9.6
Fabricated metal products, machinery and equipment	10.0	11.8	8.8	8.9	7.0	8.7	8.5	9.0	8.8

Source: Bank of Korea, Financial Statement Analysis.

Table A10.1: WAGES, PRICES, PRODUCTIVITY

	1976	1977	1978	1979	1980	1981	1982	1983	1984	1985
Wages										
Index	35.4	46.8	63.2	81.0	100.0	120.7	139.7	155.1	168.6	184.2
Rate of increase	35.5	32.1	35.0	28.3	23.4	20.7	15.8	11.0	8.7	9.3
Prices (Wholesale)										
Index	49.8	54.3	60.6	72.0	100.0	120.4	126.0	126.3	127.2	128.3
Rate of increase	12.2	9.0	11.6	18.8	38.9	20.4	4.7	0.2	0.7	0.9
Productivity										
Index	63.9	70.3	78.4	90.5	100.0	116.9	125.3	141.6	155.8	166.6
Rate of increase	7.0	10.0	11.5	15.4	10.5	16.9	7.2	13.0	10.0	6.9

Source: Data provided by Economic Planning Board.

Table A10.2: R&D EXPENDITURE BY INDUSTRY, 1983
(Million won)

	Amount	R&D/ output	R&D/ manpower
Total mining and manufacturing	348,372	0.82	40.2
Mining	2,603	0.71	68.5
Construction	2,973	0.03	18.9
Manufacturing	339,248	1.07	41.3
Food and beverage	12,587	0.44	24.8
Textiles and wearing apparel	16,869	0.56	44.3
Wood and wood products	482	0.19	24.1
Paper and printing, publishing	3,959	1.08	43.0
Chemicals	54,700	0.56	37.0
Industrial chemicals	19,299	0.97	47.8
Organic and inorganic chemicals	3,491	0.56	32.0
Fertilizer and pesticides	2,223	0.64	32.2
Synthetic resin	13,585	1.32	60.1
Other chemicals	22,683	2.07	29.9
Paints, varnishes and lacquers	4,078	2.90	25.8
Drugs and medicines	11,926	2.47	40.2
Soap, perfume and cosmetics	2,203	0.93	15.6
Other chemical products	4,476	1.89	27.5
Petroleum refineries	3,092	0.06	47.6
Oil and coal	1,651	0.26	21.4
Rubber products	7,412	1.20	50.4
Plastic products	563	0.28	20.1
Nonmetallic mineral products	6,375	0.42	25.9
Pottery, china and earthenware	268	0.86	11.7
Glass and glass products	2,555	1.40	26.6
Other nonmetallic	3,552	0.28	28.0
Basic metal	23,164	0.57	58.5
Iron and steel	17,174	0.51	66.1
Nonferrous metal	5,990	0.84	44.0
Fabricated metal products	16,495	6.11	175.5
Machinery	220,120	2.23	43.5
General machinery	14,037	1.61	23.6
Engines and turbines	8,642	1.84	22.5
Metal and wood-working machinery	273	1.03	8.3
Industrial machinery	1,017	1.58	33.9
Other machinery	4,105	1.31	27.7
Electrical and eletronic	126,672	3.44	45.4
Industrial electrical machinery	9,834	2.64	40.3
Electronic products & comm. equipment	62,453	3.40	31.9
Electrical appliances and housewares	51,055	5.11	116.6
Other electrical and electronic	3,330	0.70	22.3
Transport equipment	57,068	1.17	39.7
Shipbuilding and repairing	16,226	0.76	42.7
Motor vehicles	31,375	0.22	42.3
Motor vehicle parts and accessories	6,576	2.01	28.5
Other transport equipment	2,891	0.28	34.0
Precision machinery	5,848	4.05	41.8
Other manufacturing	992	0.82	23.6
Information and research service	675	3.66	10.7
Construction and technology service	2,873	3.52	15.8

Source: Ministry of Science and Technology

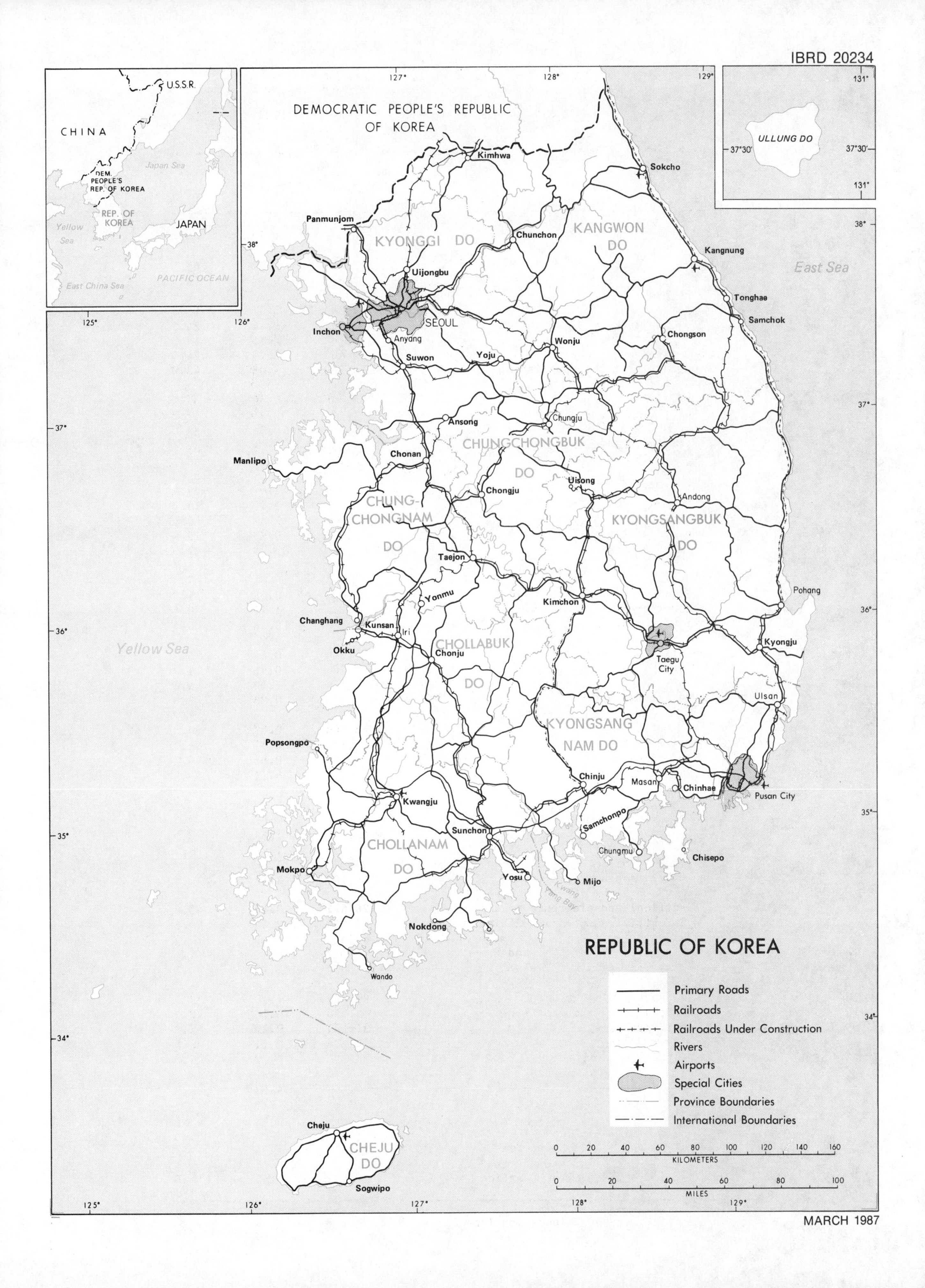
IBRD 20234
REPUBLIC OF KOREA
DEMOCRATIC PEOPLE'S REPUBLIC OF KOREA
CHINA
U.S.S.R.
DEM. PEOPLE'S REP. OF KOREA
REP. OF KOREA
JAPAN
Japan Sea
Yellow Sea
East China Sea
PACIFIC OCEAN
ULLUNG DO
East Sea
Yellow Sea
KYONGGI DO
KANGWON DO
CHUNGCHONGBUK DO
CHUNG-CHONGNAM DO
KYONGSANGBUK DO
CHOLLABUK DO
KYONGSANG NAM DO
CHOLLANAM DO
CHEJU DO
SEOUL
Kimhwa
Sokcho
Panmunjom
Chunchon
Kangnung
Uijongbu
Tonghae
Samchok
Inchon
Anyang
Chongson
Wonju
Suwon
Yoju
Chungju
Ansong
Manlipo
Chonan
Uisong
Chongju
Andong
Taejon
Yonmu
Pohang
Kimchon
Changhang
Kunsan
Iri
Okku
Chonju
Taegu City
Kyongju
Ulsan
Popsongpo
Chinju
Masan
Chinhae
Pusan City
Kwangju
Samchonpo
Sunchon
Chungmu
Chisepo
Mokpo
Yosu
Mijo
Kwang Yang Bay
Nokdong
Wando
Cheju
Sogwipo
Primary Roads
Railroads
Railroads Under Construction
Rivers
Airports
Special Cities
Province Boundaries
International Boundaries
KILOMETERS
MILES
MARCH 1987